The EVERYTHING
Home Recording Book

Dear Reader:

I became interested in home recording after my first experience in a recording studio as a young musician. When I walked into the control room, I was awestruck by all the gear. I was also dumbfounded by what things did. Eventually, I made the jump to a small home studio of my own. When I went back into larger professional studios, I watched talented engineers at work and asked many questions. Over the years, owning a home studio has helped out in so many ways. Being able to produce studio-quality demos for groups with low or no additional cost paid for the studio very quickly. The ability to record when I want, on my own schedule is so freeing. I've also, on occasion, taken home parts from professional sessions, added my own parts, saving the artist considerable time and money. It has also been a valuable practice and composition tool for myself. This book has been so much fun to write. It has satisfied my need to be a teacher, musician, and a computer technology nerd—all in one shot! I hope the way I've organized this book helps you learn what recording is about. This could have been a dry manual. Instead, interspersed with the information, you'll experience a humorous and easygoing feel. I hope you have much success in your studio! Have fun, and use your ears.

Cheers,

Marc Schonbrun

The EVERYTHING® Series

Editorial

Publishing Director	Gary M. Krebs
Managing Editor	Kate McBride
Copy Chief	Laura MacLaughlin
Acquisitions Editor	Eric M. Hall
Development Editor	Karen Johnson Jacot
Production Editor	Jamie Wielgus

Production

Production Director	Susan Beale
Production Manager	Michelle Roy Kelly
Series Designers	Daria Perreault
	Colleen Cunningham
	John Paulhus
Cover Design	Paul Beatrice
	Matt LeBlanc
Layout and Graphics	Colleen Cunningham
	Rachael Eiben
	Michelle Roy Kelly
	John Paulhus
	Daria Perreault
	Erin Ring
Series Cover Artist	Barry Littmann
Interior Illustrations	Argosy

Visit the entire Everything® Series at www.everything.com

THE
EVERYTHING®
HOME RECORDING BOOK

From 4-track to digital—all you need
to make your musical dreams a reality

Marc Schonbrun

A

Adams Media
Avon, Massachusetts

This book is dedicated to Herkimers and Chickens everywhere.

An Everything® Series Book.
Everything® and everything.com® are registered trademarks of F+W Publications, Inc.

Published by Adams Media, an F+W Publications Company
57 Littlefield Street, Avon, MA 02322 U.S.A.
www.adamsmedia.com

ISBN: 1-59337-138-1
Printed in the United States of America.

J I H G F E D C B A

Library of Congress Cataloging-in-Publication Data
Schonbrun, Marc.
The everything home recording book / Marc Schonbrun.
p.   cm.
(An Everything series book)
ISBN 1-59337-138-1
1. Sound--Recording and reproducing. 2. Sound studios. I. Title. II. Series: Everything series.

TK7881.4.S45 2004
621.389'3--dc22

2004009920

This publication is designed to provide accurate and authoritative information with regard to the subject matter covered. It is sold with the understanding that the publisher is not engaged in rendering legal, accounting, or other professional advice. If legal advice or other expert assistance is required, the services of a competent professional person should be sought.

—From a *Declaration of Principles* jointly adopted by a Committee of the American Bar Association and a Committee of Publishers and Associations

Many of the designations used by manufacturers and sellers to distinguish their products are claimed as trademarks. Where those designations appear in this book and Adams Media was aware of a trademark claim, the designations have been printed with initial capital letters.

This book is available at quantity discounts for bulk purchases.
For information, call 1-800-872-5627.

Contents

Acknowledgments

Thanks to the usual suspects . . . Mom, Dad, Bill, Trish, David, and my wonderful and patient wife Karla. James Bennett, Mark Robinson, Colin Fairbairn, and Rail Jon Rogut for expert help. To the many companies involved, this wouldn't be possible without you: Didi Dori at Waves, Chandra Lynn at Digidesign, Michael Logue at Antares, Eric and Lori Persing at Spectrasonics, Bela and the gang at Native Instruments, Brian McConnon at Steinberg, Kevin Walt at M-Audio, Bill Gardner at Wave Arts, Arjen van der Schoot at Audio Ease, Christine Wilhelmy at Emagic/Apple, Jim Cooper at MOTU, Will Shanks at Universal Audio, and Derrick Floyd at IK Multimedia.

Top Ten Reasons
to Record at Home

1. It's cheaper than going to the studio.
2. No more watching the clock waiting for the "perfect take."
3. It's a blast!
4. You can work when you want to.
5. You can produce high-quality music yourself.
6. Because you've always wanted to.
7. You can form new musical collaborations and record them.
8. Because your computer is wasting away surfing the Web— put it to work.
9. Because you can sound as good as the pros do, and you can do it yourself.
10. Certain tools exist in the home studio realm that don't exist in professional studios.

Introduction

▶ Music is so central to our lives as human beings. If you're reading this book, then making music is an important part of your life. Furthermore, whether your music is for fun or a serious interest, if you're reading this book, you would like to get into home recording. Where do you start? Maybe you've gone to a music store and come back more confused than when you walked in. All those choices! Analog vs. digital, microphones, cables, rack gear, microphone preamplifiers, mixing boards, computer interfaces, recording software, MIDI . . . The list goes on and on. Even worse, you picked up a book on home recording . . . On page six you learned about hi passing 200Hz to eliminate some rumble from a bass-heavy cardioid microphone to battle proximity effect. Proximity . . . What? Maybe you closed that book and got scared. All those terms, with so little real-world instruction on where to start and what to do.

The Everything® Home Recording Book is written for someone who has little or no experience in the field of recording. The only prerequisite is having music in your soul that you wish to record—that's all. This book is designed to take you through home recording step by step. From getting the gear, to setting up your gear and recording properly, our goal here is not to skip steps. If you're somewhat familiar with recording you might want to skip around to some of the meatier chapters later in the book. Even so, you might want to look through each chapter to make sure you haven't missed anything vital.

The field of recording is rooted in math and physics, so there's no denying the academic link and why it's used. There's no way to avoid talking about hertz and kilohertz, just as there's no way around decibels and ratios. That's because we have the daunting task of explaining sound. Warm, muddy, clear, and boomy are all terms to explain characteristics of sounds; so are 20Hz and 10kHz. Don't be scared of the math side of things; it's just one way of looking at it. You'll be pleased to find that this book explains both sides of the fence: some theory and a lot of application. No one learns this in a vacuum. No matter how many books you read, or how much physics you understand, there is no substitute for getting your hands on gear and twiddling knobs to see what happens. That's really the only way to learn. We've paid special attention to the common mistakes that beginners make. Topics like which microphones to use and setting proper input levels are covered in great detail here. While these topics might not be as glamorous as getting that "Pink Floyd" sound, these are the foundation of your recordings.

This book gives you clear information on where to start and how to sharpen your skills. But your experimentation and drive to create will teach you more than any text. I hope you're ready to learn just about "everything" about recording in home studios! Let's go!

Chapter 1

Recording Basics

The technology to record sound has been around for just a little over a hundred years. But it has come a long way since it began. This chapter covers recording history and the development of an industry that has gone from being exclusive and expensive to becoming an affordable alternative for home and semiprofessional musicians.

How It All Began

From cave paintings to the Dead Sea Scrolls, information has been written down and preserved for all to see. Recording sound has been around only since the late 1800s. Sadly, much of the history of sound itself has been lost because it occurred before the ability to record it was available. Imagine being able to hear Mozart play his own piano pieces, or to hear Abraham Lincoln speak. These memories survive only through written words and recollections of the events. Recording sounds has served not only as an important historical tool, but also as a way for music to be preserved and enjoyed.

A Brief History

In 1877 an inventor working in New Jersey single-handedly invented recording, the art and science of capturing sound. Thomas Edison recorded the tune "Mary Had a Little Lamb" on a tin cylinder and subsequently played it back. Edison's system recorded sound as indentations on a rotating tin cylinder, and the sound was then played back via a needle that felt the grooves and replayed the sound. This was the beginning of sound recording as we know it. However, tin was not a durable medium to record sound because it deteriorated upon playback. Tin was also limited to three minutes of recording time. The fidelity of the sound wasn't exactly beautiful either, but it was a start.

Edison was not the only inventor working on sound recording; he just got there first. In the late 1800s, after Edison's original invention, many inventors sought to build a better mousetrap, so to speak. Other inventors devised different disks and cylinders made of various materials to improve sound quality, recording time, and durability.

Edison

Edison pioneered the first audio recordings and brought them into people's homes. Edison's invention of the disk phonograph in 1914 was one of his most significant achievements as an inventor. His disks were more durable, produced immeasurably better sound quality, and could record longer pieces than anything else available. After mass-production of phonographs began, their price fell and allowed them to enter the homes of the common person, no longer a toy just for the wealthy. Record companies started popping up everywhere! This was the beginning of the revolution of bringing sound into the home. Unfortunately, making your own recordings was an expensive and time-consuming operation, and very few companies had the capital or the equipment to do so.

Recording Defined

What exactly is recording? Recording is the transmittal of sound waves onto a device capable of preserving and reproducing that sound. Several components are necessary to make a music recording today. First, there needs to be a sound source—this can be either an acoustic instrument or an electronic one. Then we need to get the sound source into the recording device. For acoustic instruments, a microphone is needed to convert the acoustic information into electrical signals through a cable.

FACT

The first microphone was invented in 1876 for Alexander Graham Bell's telephone system. Bell's microphone picked up sound and converted it to electricity that could be transmitted and reproduced. Chronologically, the microphone predates all recording by one year!

Electronic instruments, such as keyboards, interface directly with the recorder, bypassing the need for a microphone, although you could also use an amplifier and then record the sound conventionally. Because all electronic keyboards output their sound as an electrical signal, recording directly this way and bypassing the amplifier ensures that you get the purest signal.

Next, now that we have the sound, we need somewhere to store it. In the early days of recording, sound was etched into grooves in tin, wax, and later, vinyl. Now sound is stored in either of two ways: as an analog signal (a continuous periodic signal) or as a digital representation of an analog signal. A continuous periodic signal is like a wave in its periodic nature. Think of the waves at the beach. First comes the crest of a wave followed by a trough, then another crest, then another trough, and so on. Analog media stores the waves themselves as a continuous electrical charge. Magnetic tape is the most common analog medium. This can be the common "tape" that we all know and love, or it can be large-format, 2-inch reel-to-reel tape. Analog tape recording has been around since the 1950s and is still favored by many artists and producers for its warm, rich sound.

Digital media store a numerical representation of the wave using a code consisting of only zeroes and ones (010101010), called binary code. One way to store digital information is to store the binary information on magnetic tape. Digital audiotape, unlike traditional analog media, records only digital information—there is no sound on digital audiotape. The recording device converts analog signals to a digital representation, called encoding, on the way in and converts the digital representation to analog signals, called decoding, on the way out. These two processes are commonly called A/D and D/A conversions. When you read A/D, you say "A to D" or "analog to digital"; D/A means digital to analog.

FACT

A reel of 2-inch tape retails for about $150. But the same amount of information can be recorded to a computer hard drive or CD for a fraction of the cost.

The digital audio tape (DAT) is widely used as a high-quality stereo tape for mixing down or as a master recording. It is also possible to record multitrack on digital tape. Older professional digital multitrack systems used the same 2-inch tape that analog systems used, but encoded the tape with digital information. These systems were very expensive and out of reach for home studios. The introduction of the Alesis ADAT and the Tascam DA-88 modular digital recorders revolutionized digital recording. These machines were considerably less expensive than analog tape solutions, and, more important, they recorded on S-VHS or Super 8 videocassettes that were cheap and readily accessible. Traditional analog tape is very expensive and hard to get, more so now than ever.

The other option for recording digital information is no stranger to anyone: the computer hard drive. Hard drives store data on magnetic platters that are similar to large vinyl records, except they store information digitally. Because of the tremendous size of modern hard drives and their low cost, they are a great choice for storing music.

The format you ultimately use is unimportant; all do the same basic job of recording sounds. Each format has its own distinct advantages and disadvantages, which we discuss in later chapters.

The final step in recording is playing back the recorded sound. Both analog and digital media must convert information to audible sound waves via speakers, called monitors, or through headphones.

Early Recording Techniques

When tape recording was first becoming prominent, all recordings were done live. All the sources converged onto one track of a magnetic tape. Because there was only one track, there was no way to adjust the individual levels of the recorded instruments after the initial recording. If you didn't get the balance right the first time, you had to record the entire track again. Overdubbing, the process of adding live tracks after the initial recording, was impossible because of the mechanics of early tape recorders. As time went on, tape recorders split the width of the tape up into smaller tracks, allowing for multiple tracks. In time, recording and playing back four or eight tracks became possible.

Les Paul's Innovations

Before jazz guitarist Les Paul came on the scene, overdubbing was virtually impossible. To understand the difficulty in overdubbing at that time, you first need to understand how the analog tape machine works. In an analog tape machine, three electric heads—the record head, playback head, and erase head—handle the recording process. The record head magnetizes the tape that flows beneath it; the playback head picks up the information from the tape and sends it out to the speakers; and the erase head erases the tape when necessary. The three heads are set up one after another, so that as the record head writes, the playback head picks up tape farther along in the recording. Because each head is reading a different part of the tape, they aren't synchronized. For overdubbing to work (such as layering guitar sounds one on top of another), the artist would need to listen to the previously recorded track to know when to start, when to pick up the tempo, etc. . . . Unfortunately, since the playback head is in a different spot than the record head, the recorded signal is out of sync; it plays back later than the artist played it.

Les Paul was a very innovative man. Not only did he invent the solid body electric guitar as we know it today, he also made overdubbing and multitrack recording possible. Paul had the idea to combine the record head and the playback head into one unit, allowing artists to overdub in real time with no delay.

FACT

Les Paul's first multitrack guitar recording was the song "Lover." It featured eight tracks of guitar, overdubbed one track at a time. This was the first multitrack recording in history.

Les Paul's records were revolutionary; no one had ever heard such a thick, lush sound. Based on Paul's discovery, the company Ampex released a 4-track recorder with Sel-Sync (Selective Synchronization) in 1955. However, while this innovation made overdubs possible, most bands still recorded live and used overdubs to add solos, harmony parts, or additional vocals.

How Multitrack Changed the World

Multitrack recording was the single most important innovation in audio recording. The ability to record instruments on individual tracks, have control of separate volume levels, and add other parts after the original recording, changed the recording process forever. No longer did you have to settle for a live take. If the singer was off, you could go back and re-record individual parts. Guitar players could layer acoustic guitar backgrounds with electric guitar rhythm parts. The possibilities were endless.

The number of tracks available increased over time. At first, the 4-track was common. The Beatles, for example, recorded "Strawberry Fields Forever" on two separate 4-track tape machines, for a total of eight tracks. Modern recordings can be twenty-four, forty-eight, or, in the case of computers, several hundred tracks. For you, the home-based musician, this process allows you to slowly build up arrangements one track at a time. You can start with a bass line, add a guitar part later, track some vocals later—all by your lonesome. The finished product will sound like one large, live band even though you played it all yourself.

But home studio owners aren't the only ones who work this way; Trent Reznor of Nine Inch Nails always multitracks. He records alone in his home studio, multitracking to build songs. Tom Scholz of the group Boston records the same way, playing each instrument one at a time.

Modern-Day Developments

We live in the digital age. Everywhere around us technology is changing the way we work, play, and communicate. The computer has become a fixture in the home, and it's hard to imagine life without one. The need to create coupled with advancements in technology are allowing even the average hobbyist the chance to create and share quality music without going into considerable debt.

The Advancement of Technology

Analog multitrack recorders capable of recording twenty-four or more tracks can cost a lot of money. Even now, though they are less popular, it is easy to spend $30,000 to $50,000 on a good one. Their prohibitive cost meant that for a while, home studios were available only to rich, successful musicians. Digital technology has brought the cost down considerably. Digital tape machines such as the ADAT, while not cheap, are nowhere near as expensive as multitrack analog tape machines. When they were introduced in 1992, Alesis ADATs went for around $3,500. These modular tape machines started showing up in professional studios, and more and more home studios were getting equipped with digital recorders.

ADAT records on standard S-VHS videotape. The tapes are inexpensive and easy to transport. The design of the tape makes it durable, unlike analog tape that deteriorates with handling.

As the technology and time advanced, recording studios turned to the personal computer. Digital audio can be stored on its internal hard drives, and the monitor and mouse take editing to a whole new level. Software provides an interface for laying out tracks and editing them visually in ways that were never possible in the analog or digital tape world. As we discuss in later chapters, digital audio uses nonlinear technology. This means that the audio is free to be placed anywhere in time, unlike a tape-based machine on which you record at a specific point in the tape. Unless you cut out that section of tape and splice it somewhere else, you can't move things around with analog tape. But with digital audio, moving audio is as easy as pointing and clicking.

Using computers in studios came with its own problems: The computers themselves were not able to handle the tremendous strain that digital audio required. To a computer's brain (the central processing unit, or CPU), digital audio is very complex to work with. The addition of signal processing was too much for the computers of the late 1980s and early 1990s to handle. The solution was to use add-on cards inside the computer to help process the digital audio signal. One of the most successful products is Digidesign's Pro Tools. Pro Tools uses a combination of hardware to perform digital signal processing (DSP) and software to arrange music. Professional Pro Tools and other systems like it are still very expensive. It's easy to spend $30,000 to $50,000 on a nice Pro Tools rig. Pro Tools was one of the first proprietary systems available, a combination of software and hardware for recording music in a computer. Today Pro Tools is the standard in recording studios around the world. Other systems exist today, but none with the popularity and compatibility of Pro Tools.

How Technology Made the Home Studio Possible

The home studio has followed a path similar to that of professional recording studios. In 1979, Tascam invented the Portastudio, a 4-track recorder that used standard audiotapes. It was priced around $1,000, which was very inexpensive for a unit of its type. It caused a revolution, and in one step created the home studio market. The unit was small and compact and could be taken anywhere. Four tracks could be recorded and mixed separately in the unit and later mixed down to a final stereo cassette. Musicians quickly began using the Portastudio for creating their own music and making demos. The Portastudio line by Tascam is still popular today and comes in many shapes and sizes, both digital and analog.

FACT

Recording signals come in two forms: monophonic and stereophonic. A monophonic signal can be reproduced using only one speaker. Old radios with one speaker are monophonic. All of the modern music we listen to is mixed for stereophonic sound, which uses two speakers: left and right.

In the digital world, in the 1990s, the hard disk began showing up as part of standalone recorders, greatly increasing the quality of recorded sound. Because hard disks were able to hold more data, they became a viable solution to storing digital audio. Digital audio is very large: each monophonic track takes 5 megabytes of memory per minute. A typical ten-minute song consisting of eight tracks requires 400 megabytes of storage space. By today's standards that's not very much, but in the early 1990s most home computers shipped with 500-megabyte drives, total! As the computer grew in popularity and power, it became feasible for a computer with a simple audio interface to handle the demands of digital audio without the need for additional DSP cards. Computer recording software such as Cubase, Digital Performer, Sonar and Logic answered the call by providing MIDI (musical instrument digital interface) and digital audio in one package. Computer recording software is immensely popular because of its ease of use, relatively low cost, and the power of what you can achieve with just your home computer.

What does all this history mean for you and your home studio? Being able to layer track upon track is a critical part of the home studio experience, especially if you work alone. Many bands record albums one layer at a time for greater control.

Elements of a Professional Recording

As a home studio owner, or soon to be one, you should be aware of how the professional studios operate and what techniques they employ. In the end we are all trying to do the same thing: get sound onto a recording device, spice it up, and mix it to a final product. We all want to get the best sound possible. The differences in techniques directly affect the quality of the final product.

ALERT!

Professional recording studios can charge up to several thousand dollars an hour for recording services! For the cost of one session in a professional studio, you could take that money and invest it in your own studio and work whenever you want to.

Why Your Favorite Recordings Sound So Good

Cue up your favorite recording, one that you think is recorded well. Sit back and listen closely. Notice how all the instruments blend together well, how no instrument sticks out of the mix more than it should. Notice how you hear virtually no background noise. All the instruments sound present, the drums don't sound far away, and it sounds as if you're in the same room as the band. The recording has a smooth and polished sound to it, no harshness to your ears. These are all qualities of good engineering, good mixing, and good mastering.

When you listen to a professional recording, realize that you are listening to months, if not years, of hard work recording and mixing the music. Big studios also have access to the finest equipment, the best microphones, acoustically perfect rooms, and most important of all, experienced engineers to run the sessions. Does this mean that your home studio masterpiece will sound bad? No, not at all! With some basic equipment, a little knowledge, and your inspired music, you can make professional-sounding recordings. Recording sessions are broken up into three main components: preproduction, production and engineering, and postproduction.

Preproduction

Preproduction involves everything that comes before the actual recording session. This can include selecting the right material to record, rehearsing the band, and getting ready for the recording sessions. For the home studio owner, this involves working out your material so that you can record it. It also might include purchasing gear to facilitate a particular project, such as buying a second vocal microphone to record a vocal duet for a new song. Basically, anything that you can do in advance to make your recordings go more smoothly is preproduction.

Production and Engineering

Production involves the actual recording sessions. At the sessions, the engineer runs the recording show. It's up to the engineer and any assistants he or she might have to set up all the microphones, place the microphones for optimum sound, get proper recording levels, run the mixing board, operate the recording device, and make sure everything sounds good. The engineer is the most important link in the chain (besides the musicians themselves) in getting a great-sounding recording. Engineering, like any other skill, requires a certain level of artistry and practice for proficiency to improve. An experienced engineer will be able to identify problems and quickly find solutions.

QUESTION?

How can I learn more about studio engineering?
Many colleges offer courses in recording techniques. Check around the colleges in your area to see if they have anything that fits your schedule. You could also volunteer your services at a local studio in order to gain experience and learn the business from the inside.

Editing and overdubbing might take place in subsequent sessions, but it's still considered production. In your studio, you will most likely be wearing all of the various hats needed to make a recording. It will be up to you to properly set up your equipment and the microphones, run the recording device, and engineer the recording. This can be a tall order to do all at once, but the chapters to follow will show you how to get started easily. With a little practice you'll be off and running!

Postproduction

Postproduction includes anything that happens after the recording sessions. Most often, postproduction involves mixing the tracks to a polished uniform sound. Mixing involves several key elements:

- **Track levels:** Loudness of each track
- **Panning:** Side-to-side placement in the mix
- **Equalization:** Boosting or cutting certain frequencies in the mix
- **Effects:** Adding signal processing such as reverb, delay, and compression in order to achieve a polished sound
- **Mix down:** Mixing all the tracks into a single stereo pair suitable for distribution or mastering

Even the most basic studio has the capabilities to do all these things. Remember that the basic sequence of events is always the same: sound capture, recording, and playback. Now that we explained a little about the history of the recording process and got you thinking about some concepts, it's time to shift gears and move into your home studio to find out what you need to get started.

Chapter 2

So You Want to Cut a Record . . .

You're ready to make the leap from weekend warrior to a home recording studio owner . . . but how do you do it? What do you need? It's easy to become overwhelmed with all the choices when you're getting started. This chapter serves as a primer and guide to getting you started on your home recording odyssey. Let your creative juices flow!

Bringing It Home

So just what *can* you do in a home studio? What is it, exactly, that you will be able to do once you have your studio set up? Even if you have no experience in this, you'll find that your natural talent and years of listening to well-produced music have given you more tools than you thought you had. It's all about listening.

Recording

The most obvious thing you can do in a home studio is to record sound onto tape or disk. Whether you are learning to use microphones effectively or just plugging a keyboard in, recording covers the whole spectrum of capturing sound. For those who have never recorded before, the process can be very rewarding. Having the ability to come home from work and spend a few hours in your studio creating music is very freeing, and the icing on the cake is that you have total control. Whether you are making an elaborate multitrack masterpiece or simply singing and playing guitar, you need to know how to get a good sound. This is where you learn the basics of engineering: setting levels, choosing and placing the correct microphones for the best sound, and dealing with mixing. For the keyboard players out there, using MIDI (an electronic standard used for the transmission of digital music) is an important part of the recording process. All these elements fall under the umbrella of recording.

Mixing

Mixing is generally done after the initial recording sessions. Mixing is the art of setting the loudness and sound color of each instrument that you record. Mixing is a learned art, and like any other skill, it takes practice. Using faders to control the volume of tracks helps the instruments sound more cohesive and balanced. Equalization, which we will refer to as EQ, is the boosting or lowering of certain frequencies. EQ can transform a muddy bass into a clear sound, or turn a thin, lifeless guitar into a round, warm sound.

Effects are also a major component of mixing. Effects can help you achieve a more polished final sound. For instance, you can add reverb to give the illusion of having recorded in a large space or use compression to even out sudden changes in volume.

All of the things discussed here will be covered in more detail later in this book. Don't be worried if it sounds like a lot to learn! Chapter 10 discusses mixing in more detail, and effects are covered in Chapter 14. MIDI is discussed in Chapter 7.

Editing

Editing can be considered cut, copy, paste, and whiteout for sound. In the world of editing you can cut a section and re-record it; you can go back and fix a note that sounds bad.

The most powerful editing options lie in the computer. Standalone, all-in-one recording units are capable of editing, but computer software takes it to a much higher level. Imagine that you are recording a song with a structure of chorus, solo, chorus. Suppose the choruses are exactly the same, and when you recorded the piece, the second chorus sounds better. Computer software gives you the ability to cut the first chorus and copy the second in its place. You don't even have to play repeat sections twice; you can just loop together small repeating bars of music. You can even click and drag sections of music around with your mouse. Did you decide after the fact that you want to record an introduction to your masterpiece? Just drag the original audio to the right, make room for the new, and paste it all together. This is just the tip of a very large iceberg . . . So now how do you get started? First, you've got to determine what your needs are.

Assessing Your Needs

You've scoured the Internet. You get every music gear catalog known to mankind. You've been to the local music store countless times. You know it's time to start doing some home recording, but the myriad of choices and lack of concrete how-to instructions is getting to you. Have no fear! You're in the right place now.

"Home recording" is a broad term—musicians have different needs and ideas about what the studio is going to do for them. For some, home recording is a sketchpad for small ideas that later might be taken to a professional studio. For others, their home studios are used to flesh out ideas so they can present what they have in mind to members of their bands. Still others use home recording as a way to save money. If you plan to make frequent demo recordings or just want the flexibility to work whenever you want, your money is better spent investing in a home studio of your own, because professional recording studios can be very costly. And finally, some musicians use their home studio as a creative tool to write, produce, and ultimately sell their own music. Some home studio owners enjoy the process so much that they eventually upgrade their equipment and open their studios to the public. What you do with your studio is as personal as the music you create. The sky is the limit and, with modern technology at your side, you will be armed with all the tools to make great sounding recordings.

Start Simple

It's very easy to go overboard in this field. There is plenty of equipment out there that you could spend lots of money on. Everyone fantasizes about the professional studio with a 10-foot-wide mixing board and floor-to-ceiling rack equipment. Some musicians do need all that stuff, but what do *you* need? The first step is to assess your needs. Ask yourself these questions before you start buying gear:

- How many instruments do you need to record at the same time?
- Are the instruments electric or acoustic?
- Will you use MIDI or sequenced instruments?
- Do you want to use your computer for any part of the recording process?
- Do you need portability?
- Do you plan to distribute or sell the music recordings you make?

Keep the answers to these questions in the back of your mind as you read this book. The theme you'll see repeated throughout is: Make the most of what you have. Expensive gear won't necessarily make anything better. What really matters is what you do with what you have. Too many studio owners get caught in the trap of having the nicest toys without understanding or utilizing their gear to the fullest. Imagine Grandma driving a Ferrari to church once a week. Bit of a waste, eh?

FACT

The recordings of the 1950s and 1960s were recorded with equipment that would be considered limiting nowadays. You'd be amazed at how many major recordings that you know and love were done by a highly skilled recording engineer on very basic equipment. Rudy Van Gelder's 1950s and 1960s Blue Note jazz sessions come to mind, as do George Martin and the Beatles.

Have a Goal

Having a goal seems like a simple idea. However, many people jump on the home studio bandwagon without even considering a goal. Ask yourself, "What do I want to do with a home studio?" The result you're looking for—be it demo tapes to send to local clubs or recordings to sell after a show—will help you determine what you need in a home studio. More often than not, at first you'll want to start small. You can always upgrade as you become more skilled at the process. Remember, home recording is a skill like any other, and it takes a while to get really good at it.

Seek Advice from Others

More than likely, you know other musicians. It's safe to assume that a percentage of them will also own home studios, in one form or another. Spend some time talking with them and, if possible, get hands-on demonstrations of the equipment they use. Find out how they use it and listen to how their final product sounds. Another invaluable resource is your local music store. Many of these stores are staffed with very talented musicians. Many of them, in addition to knowing a great deal about the equipment they sell, have home studios of their own. Ask them what they use, and what they use it for. They might even let you listen to a CD of their work. There are also many magazines and online sources devoted to recording.

Popular recording publications include:

- *Mix Magazine*
- *EQ Magazine*
- *Recording Magazine*

In the online world, you'll find many resources and bulletin boards where you can exchange ideas with other home studio owners. Many professional engineers frequent these sites and you can learn a lot by listening and posting. For general recording tips, check these Web sites:

- *www.homerecording.com*
- *www.harmonycentral.com*
- *www.audioforums.com*
- *www.recording.org*
- *www.tapeop.com*

For those of you who still use the Usenet online message system, here is a sample of some very active, useful groups.

- alt.music.4-track
- alt.music.producer
- rec.audio.pro

Shopping for Gear

Okay. Now you're really excited. You're ready to start. You've thought through the whole process. You've talked to other studio owners. You've looked around on the Web. Now it's time to start shopping.

Depending on where you live, you might have access to music stores that carry a lot of recording equipment. If you do live near one, your best bet is to buy from a store, instead of online. There are a bunch of good reasons you should buy from a store. For starters, you'll get to see, touch, and even use some of the gear you are planning to purchase before you spend a dime. You'll also be able to get advice from the salesperson on what might suit you best. Developing rapport with an individual salesperson is important because, as you keep going back to the same person for all your gear, not only do you develop a nice business relationship (which might result in discounts), you might also get honest, real-world advice.

But there's nothing wrong with buying gear online, either. If you live in a remote area, this might be one of your only options. If you purchase online, returning gear you don't like is a pain, and will always be that way. But then again, there are some great deals online. If you shop around, you can get a great price. Many stores also price-match, so even if you've found a better deal somewhere else, bring the information to the attention of your salesperson. Many stores will accommodate you in an effort to keep you as a loyal customer. You can also find some great deals in the used-gear market; check out eBay.

Don't feel pressured into buying more than you need at first. You can always upgrade as you go. Certain components don't change from setup to setup, such as cables and microphone stands, so you won't waste money upgrading those items.

Creating a Budget

Here comes everyone's least favorite subject . . . spending money. You work hard for it, and the last thing you want to do is squander it on equipment that isn't suitable for you, or doesn't get the job done. The good news is that there is a studio to be had at almost every price level, and you can get started with a basic studio for around $250, maybe even less, depending on your configuration and what you might already own. The bad news is that there's a lot more gear available, and you can easily get carried away and spend many thousands of dollars on all the various gear out there. Figure out exactly what you can spend at first.

Your budget should take into consideration the following points:

- What is the maximum you can spend?
- Do you want the ability to record more than one track at a time?
- How many microphones do you need?

- How many interconnecting cables does your setup require?
- Do you need computer recording software?
- Do you need a computer recording interface?
- Does your computer need to be upgraded to handle the demands of working with large music files?
- What signal effects do you need?
- Do you need a separate mixer? (This is becoming less necessary these days.)
- How do you plan on listening to your work (headphones or speakers)?

As you can see, this is an important list. You must take all these points into consideration when planning your budget. Since every studio is different, this book talks about general setups, and you can modify the setup that is closest to your needs.

What You Will Really Spend

You've already learned in Chapter 1 the three elements necessary for recording sound: something to capture the sound, something to store it and play it back, and something to hear it played back. First, you need a sound and a device capable of capturing that sound, usually a microphone. Some instruments interface directly via cables; keyboards and amplifier line-out jacks are examples of direct instruments. Next, you need a recorder capable of recording the sound and playing it back at a later time. Last, you need something to hear the recording with—either speakers or headphones. These elements are commonly found in all studios, regardless of price, quality, or use, hobbyist or professional.

For sound input, the typical equipment is microphones. Microphones range from around $60 for a starter variety, $200 for a good one, and $500 and above for some really top-of-the-line microphones. You can even spend several thousands of dollars. Figure out how many you think you'll need. You can probably get away with one for each instrument and two to four for a drum kit, depending on how you set up the kit. If you are recording one instrument at a time, you can get away with fewer microphones. You might pool the money you save by doing this and buy one or two higher quality microphones. Instruments such as keyboards, drum machines, and guitar effects processors can plug directly into a recording source. Also, many amplifiers feature direct outputs and bypass the need for a microphone; all you need is a cable. Cables are cheap, thank goodness! You might need a direct box to change the impedance of certain instruments to match the input of the recorder. We'll discuss that more in Chapter 11. Direct boxes range from $30 to $100.

For recorders, you can start at the low end and find a cassette tape, 4-track recorder for about $100. The high end of recorders can exceed six figures, but there are many great-sounding devices in the $300 to $1,000 range. To record with a computer, you'll need software that ranges from free to $1,000. You'll also need a computer interface that accepts audio and possibly MIDI if you plan to use that. Depending on how many sources you need to record at once, computer audio interfaces can range from $100 to more than $1,000. MIDI interfaces are less expensive, and you pay more depending on how many MIDI inputs, or separate instruments, you need to use at the same time. Expect to spend between $50 for a simple one-input/one-output MIDI interface and up to $400 for eight MIDI devices.

The spectrum of home studios can be broad. How much you need to spend depends on what your goals are. It's probably best to start at the low end and expand as you're learning.

In order to play back the recorded sound, most home studio owners start with a pair of decent headphones. Headphones range from $30 to $200. If you choose to use professional speakers, called "monitor speakers," you can expect to pay anywhere from $100 to $800 or more for a set. Some monitors are self-powered and don't require additional amplifiers to run; others need an amplifier, which will cost you money as well! Your best bet is to go for self-powered monitors. There are some great ones for $200 to $300. A low-tech solution is to monitor through your home stereo. It's not the optimal way to go, but it might tide you over until you can afford more.

How Not to Get Carried Away

It's so easy to get carried away in a music store. You go in for one thing, walk out with five things you didn't need. This is known as G.A.S. (gear acquisition syndrome). This ailment affects many musicians who fall victim to the grandeur of a music store that has "everything." Out of the three elements of your studio, it's important to balance the quality of each part. The result is only as good as all the equipment you use. Add one weak link and the chain will break. For example, if you blow all your cash on a state-of-the-art digital recorder, and you plug a cheap, noisy microphone into it, the digital recorder will play back a noisy signal, in perfect digital quality. See the problem? We address the issue of G.I.G.O. (garbage in, garbage out) later in this book. Go into this process knowing what you need to accomplish and what means you have to accomplish it with. Try your best to be even and fair with the quality of the components you choose.

Typical Setups

Let's take a look at some typical setups for various types of recording systems so you can get an idea of what equipment is commonly used. Chapter 6 goes into much greater detail on this subject, but for now you can get an idea of how people work.

For Working Alone

If you work alone, you are almost definitely going to want a multitrack recorder. The ability to layer track upon track affords you the ability to create complex arrangements on your own. On the low end, a cassette tape–based recorder can get you started for a very low price. Many people who are new to recording opt for this method to get used to working. You can choose a standalone, all-in-one studio, such as the Roland VS series or one of the many systems available from Tascam, Fostex, Roland, and Yamaha. These studios contain everything you need to get started: multitrack capability, internal effects, faders for mixing, EQ, and, in many cases, built-in CD burners.

On the computer side, many people find that the editing power of computer software makes the computer a very attractive choice. Recording technology is headed toward the computer at this point. Standalone units will always be around, but as the years go on, they are starting to resemble computers more and more. You can even get a standalone recorder with a monitor, mouse, and keyboard jack in the back. So what's the difference? A studio-in-a-box comes with some guarantees. If the box states that it can record eight tracks at the same time and play back sixty-four, then it's going to do that without a glitch. If you use a computer, the number of tracks you can create is limited by the power of your machine. The more powerful your machine, and the faster your disk, the more you can do. Check out Chapter 5 to learn how to make your computer a fire-breathing digital audio monster.

Many "solo" engineers/players own just a few microphones, usually one all-purpose and one specialized microphone. A lot depends on what instruments you plan to record. Many studios involve MIDI to control drum machines and keyboards. If MIDI is involved, you'll find a computer in the studio for sure. While there are standalone sequencers, they are rarely used anymore. The computer is far more prevalent. The solo home studio owner doesn't require huge amounts of space, and usually a corner of a room or a desk area is enough to get anyone started. Since the computer is so popular, many home studios are found around computer workstations and desks.

For Working in Groups

If you're in a group, or just plan to record a lot at once, you have some choices on how to proceed. For live groups, multitrack recording isn't a necessity, although it's nice to have. A good-quality stereo recorder, such as a professional cassette tape recorder, a digital audio tape recorder (DAT), or a standalone audio CD recorder, can do a great job. Some groups even use a minidisc recorder with a stereo microphone placed in the center of the group to get above-average results. With the exception of the minidisc setup, all these scenarios require a mixer. **FIGURE 2-1** illustrates what a typical mixer looks like.

FIGURE 2-1   ▼ Mixing board

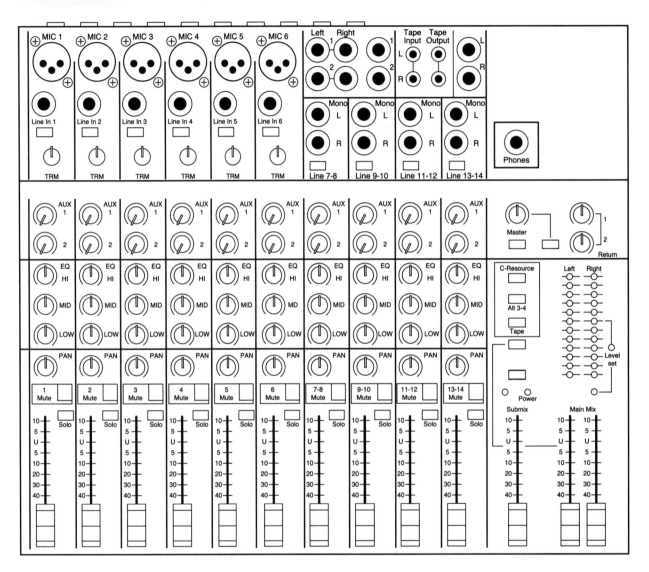

A mixer allows several sources of sound to come together and be mixed together into one stereo output. You can take eight or more microphones and mix them together into one stereo sound. The stereo output is then attached to one of the recorders—cassette tape, digital audio tape, or standalone audio CD. The nice part about this system is that it's not all that expensive; however, there are some serious drawbacks to it. First of all, the balance of the group has to be set in the mixer before the recording takes place. Since you're not multitrack recording, what you get on the final tape is all you have. Also, you have very little ability to add individual effects, except again through the mixer at the time of recording. If you mix well and set up the effects well you can get a good sound this way, but it's very difficult. If, after you're done, you realize the snare drum is too loud, there's little you can do. Even so, you'd be surprised to know how many albums have been recorded this way, especially jazz records.

Those who step up to multitrack recording do so in much the same way that the solo artist does. However, there are specific concerns that need to be addressed. How many instruments are going to recorded at once? In the case of the standalone, all-in-one studios and computer setups, the number of simultaneous inputs is crucial to achieve the ability to mix the sounds after the fact. By placing individual instruments on individual tracks, you have greater control over their relative sound and volume levels. When you are limited to a few inputs, you have no choice but to place multiple instruments on the same track, losing the ability to balance them after you record. Typically these setups use a lot of microphones. You'll need one input for every microphone you use. Your needs as a multitracker really depend on what you're recording and how much control you want.

Portable Setups

If you are doing a lot of your recording at gigs, you'll need a setup that is portable and easily mobile. Minidisc, DAT, and the standalone, self-contained studios are great for this. If your live gig has a soundperson, you can benefit from his or her gear as well. You can get a stereo mix from the soundperson and plug into a DAT or CD recorder and you are good to go. (That is, of course, assuming that the mix off the board sounds good.)

Until recently, good computers haven't been very portable. This fact is changing. In recent years, laptop computers have become extremely powerful and can handle live recordings and even many simultaneous tracks. These computer setups come at a hefty price tag, however. The combination of state-of-the-art computers and interfaces to get audio into the computer put this setup out of the reach of many. But hardware manufacturers are addressing this concern. Home studio gear is getting smaller and more powerful and will most likely continue to do so.

What's been missing in all this discussion about gear, options, and budgets is the creative spark. That is the spark that only you can provide. Recordings can't make things magically appear. No matter what kind of gear you have, from a $100 4-track, to the most decked-out computer rig with all the fixings, if you don't bring your creativity into play, nothing happens. We've all experienced good players playing cheap instruments and still sounding great. We've also experienced amateurs playing expensive gear and sounding terrible. Keep that in mind as you go through this book. You make it happen.

Chapter 3

E Elements of a Home Studio

Now that you understand the basics of recording, it's time to explore in greater detail the elements that go into a studio. As you start to make choices about what to include in your studio, you want to understand the equipment well so you can make informed decisions, and get past the advertising rhetoric.

The Center of It All—The Recorder

The center of any studio is the recording device. No matter what you choose—computer or standalone—all the magic happens at the recorder. Picking the right one for you and your needs is important, so be sure to do your homework and get it right.

Using a Computer

If you own a computer already, especially one that's recent, you might own a recording device just waiting to serve you. (Chapter 6 covers computer setups in much greater detail.) Why would you want to use a computer? Well, for one thing, if you already have one, there's less to buy. At this point, with the exception of the cassette-based, 4-track recorder (which is now on its last legs), the alternatives to computers all utilize digital technology just like a computer.

FACT

Digital recordings are stored as binary information. Music going into the recorder goes through a complicated analog-to-digital conversion that turns sound into binary information for storage.

So what kind of computer makes a good recording computer? The difficulty in answering that question lies in the fact that technology changes so quickly. What was current six months ago is considered old news today. Computers are divided into two main types by the operating system they use: Microsoft Windows or Apple Mac. Many people swear by whichever they use, and great music can be made on each. Traditionally, professional music studios have relied on Apple computers, but Windows-based systems are becoming more and more popular, especially in home studios.

The computer debate can get very silly and many people get carried away with numbers and current trends. The basic rule of thumb is age: If your computer is less than two years old, you'll be in good shape. The problem is that, while current computers work well now, they do so only because the current software is optimized for the current technology. When newer, faster computers arrive, the software manufacturers change their software to work with the enhancements of the new chips. When you try to run new software on old chips, you can run into problems.

ALERT!

It's important to buy the correct software for your machine. Every software manufacturer has minimum system requirements; it's up to you, the educated consumer, to make sure your machine qualifies. Software is usually not returnable.

What's So Great about a Computer Anyway?

There's nothing "great" about computers. Computers are simply tools that let you get a job done. Let's talk about the pros and cons of using a computer for music.

On the pro side, here are several advantages to recording music with a computer:

- You might already own a suitable machine.
- Computers have increased sound quality.
- You can edit on a large screen using a mouse.
- You can take advantage of powerful music software.
- It's easy to turn your music into Mp3s and share them online.
- It's easy to burn and label CDs of your music.

On the con side, here are several disadvantages to recording music with a computer:

- Computers crash (and often at the worst possible times).
- Except for laptops, you can't take computers anywhere.
- The price of software and hardware can be steep.
- New software has a learning curve.
- The pace at which technology changes can be annoying.
- Computers are susceptible to viruses.

If You Don't Have a Computer

So what if you don't have a computer? Or maybe you have one, but it's impossible to get your family away from it at the times you want to work. If you fall into one of these categories, you still have many options. There are plenty of recorders that stand alone. The playing field consists of cassette-based 4- and 8-track recorders, digital hard-disk recorders, digital multitrack tape machines, and studio-in-a-box solutions. The standalone solutions have portability going for them. You can easily take a standalone recorder with you wherever you go. Even the rack-mounted Alesis ADATs and the Tascam DA-88s can be thrown into a rack bag and taken to sessions. While the editing is superior on a computer, more and more of these recorders have the ability to interface with a computer later on to enable complex editing and sound manipulation, giving you the best of both worlds. If you already have a mixer and outboard rack effects, a standalone unit is a great idea for you.

The studio-in-a-box is another type of standalone recorder. It features inputs, some microphone preamps, recording, integrated effects, integrated mixing, mastering, and many include CD burners so you can burn your final product. If you're just starting out and don't own any mixers or effects, these studios-in-a-box are a really great way to go.

FACT

The all-in-one studio has brought digital studio quality to the masses. Before the advent of digital recording technology, cassette tapes lacked the sound quality and fidelity of professional studios. But now anyone can make a great-sounding recording, without spending a fortune.

Roland was one of the first companies to introduce the concept of everything under one roof. The VS-880 was a revolutionary product because everything was done in one small tabletop unit. It was small, portable, and, considering all you got in one package, surprisingly affordable. Today there are lots of models and manufacturers to choose from, and many price points, too.

There are some downsides to these units. The number of tracks you can record at the same time can be limited. You need to make sure it can handle what you have in mind for it. The quality of the effects is a limitation—whatever you get you get—and they can't usually be upgraded. Some units allow you to use external effects through an insert jack, but not all of them do. Many units of this type store music on either a hard drive or, more recently, on compact flash or smart media cards. When the drive is full, you have to stop recording. Some devices let you burn the data tracks as a backup, and some don't; you have to mix, finish, and delete before you can do anything else. Disadvantages aside, these units are very popular and sell very well. Anyone interested in home recording would be wise to consider what models and features are available.

Capturing Sound

The next step is getting sound into the recording device. Sound gets into a recorder either by being plugged directly into the recorder with a cable, or by using a microphone to pick up the sound.

Microphones

If you play an acoustic instrument, a microphone is necessary to convert the sound waves into electrical signals that can be recorded. What makes any microphone different from any other microphone? This is more important than just hearing "use an SM-57 on guitars," or "use a beta-58 on vocals." As a budding sound engineer, you should understand why engineers choose different microphones for different purposes.

The simplest answer is that every microphone hears sound differently. Some microphones only hear what's directly in front of them (unidirectional). Other microphones pick up everything from all sides (omnidirectional). And others hear sound on two distinct sides (called figure eight). If you were trying to use one microphone to record a room full of sound coming from all sides, you would choose a microphone that hears sounds from all sides (the omnidirectional microphone). If you were trying to zero in on just one instrument, and trying hard not to pick up other sounds, you would want to use a unidirectional microphone pointed right at the sound source.

Knowing what direction a microphone hears is usually the determining factor in choosing which microphone to use. You must choose the one that works best for what you're trying to record.

Some microphones also hear certain frequencies of sound better than others. This is called frequency response. You wouldn't want to use a microphone that can't hear very high signals clearly on a high-pitched instrument, or low signals on a low-pitched one. Microphones also color the sound they transmit; this is tied into frequency response. These colorations are what give microphones their distinct sound—"warm," "clean," or "clear" are terms commonly used to describe sound colorations. Because of construction differences between microphone manufacturers, every model sounds unique. With some training and experience, you'll be able to pick the right microphone(s) for your setup. Getting the right type of microphone is more important than buying a particular brand. Chapter 9 gets into all the nitty-gritty details of microphones.

Microphone Preamps

Microphones don't produce much signal by themselves. If you plugged one directly into a recorder, the sound level wouldn't be very high. You could try to compensate by turning up the entire track, but unfortunately you'd turn everything up, including the noise. What you need is a preamplifier. The preamplifier raises the output of the microphone loud enough to make the signal clear and strong.

How many microphone preamps will I need?
You will need one preamp for each microphone you record at the same time. If you don't need to record live groups, you might need only a few microphones, recording one or two tracks at a time.

Commonly, preamps are built into mixing boards, studios-in-a-box, and computer interfaces, so you might not need to purchase these separately. You can also buy individual microphone preamps if your device has none or you need more. The number of microphone preamps that your mixer or other recording device has is a critical factor in determining if the gear is right for you. This is the time for you to evaluate how many microphones you plan to use at once.

Direct Inputs

Any instrument that plugs directly into the recorder is called a direct input. No microphones are necessary in these cases. Keyboards, synthesizers, drum machines, and certain guitar and bass amplifiers are equipped with "line outs" that can be plugged directly into a mixer or a recording device. A guitar or bass plugged in directly via a cable won't be "line level"; that is, its signal won't have enough "juice" to be heard. It also has the wrong impedance, which is an electronics term referring to how much "force" the signal has due to how it impedes the flow of electricity out. Guitar and bass are very high impedance sources and generate very low output levels. Even though guitars and basses appear to have line outs, you'll need a direct box, or a "DI," if you want to record a guitar or bass directly without an amp. A direct box simply takes a signal that is too low, is too loud, has the wrong impedance, or is an unbalanced signal, and converts it to a perfectly balanced line-level output.

FACT

Direct boxes come in two flavors: passive and active. Passive DIs don't require extra power to run; active DIs do. Active direct boxes can offer additional signal strength for very low output instruments. In most cases, passive direct boxes will do just fine. Every studio should have at least one direct box.

Another nice benefit of a direct box is that it can isolate nasty buzzes that studios encounter from time to time via a button called a ground lift. Even if your amplifier or keyboard has a line out, if you are getting buzzing that's driving you crazy, try a direct box between the output and the recorder's input, and flip the ground switch. Many times this will do the trick. Anyone who has played on stage has encountered a direct box; it's indispensable. If you plan to record bass, guitar, or certain keyboards that aren't line level, you will need a few direct boxes.

To Mix or Not to Mix

No piece of gear is more closely associated with the recording studio than the mixing board. Historically that's where the engineer spends most of his or her time working. A mixing board is simply a device that takes many individual audio channels and mixes them to a stereo left and right output. Mixing boards also allow you to preamplify microphones and adjust EQ. Despite all this, mixing boards are no longer a necessity for everyone. But who needs one? And how do you benefit from having one?

Why You Might Want a Mixer

If you have a standalone recorder, you need a mixer. Standalone recorders only record and play back; they don't set levels, they don't provide EQ, they just capture sound. So you need a mixer to control audio levels. This is why the mixer is so closely associated with the recording studio. It's only in recent years with the studio-in-a-box and computer software that mixerless setups are possible. Traditionally every recording device was a simple record and playback machine; the mixer wasn't optional. Even as computer systems become more commonplace, some engineers still like having the control of a mixer in their hands, instead of mixing with a mouse.

ALERT!

If you combine two overhead microphones and the snare drum microphone into one channel, you lose the ability to control the volume of the individual signals. If you're going to do this, be *very* careful to get a great-sounding mix of those channels before you record anything for real. Record those tracks as trials to get the best sound before you commit to anything, because once it's on there, you can't change it.

If you've had experience with a mixer in the past and you feel comfortable with it, you might want to use one. You might also want a mixer if your recording device has limited inputs. Suppose that you have an 8-track recording device but the maximum you can record is eight channels. If you're recording a live band that uses more than eight signals, you might want a mixer to pare down some of the signals. For instance, instead of a drum kit using four or more tracks of the mix, you can use a mixer for the drum tracks. You can take eight microphones on the drums and mix them down so they take up fewer inputs. The only downside to this is that whatever gets mixed into one channel is there for good.

Why You Might Not Want a Mixer

If you own a studio-in-a-box like the Roland VS (Virtual Studio) series or any of the comparable products by other manufacturers, you might not need a mixer: They contain inputs and mixerlike volume controls, EQ, and everything else you need for each track. All of the tape-based, 4- and 8-track recorders allow for mixing and EQ, so you won't need a mixer there. If you're going the route of the computer-based studio, you might not want a mixer because you've already got the flexibility that a computer's virtual mixer gives you. The choice is up to you.

Achieving Portability

The terms "recording studios" and "portable" have rarely been used in the same sentence in the past. But with the miniaturization of technology, studios can be tucked under your arm and taken wherever you want. This is also great news for those with small apartments where space is at a premium.

Laptops

In the last few years the laptop has gone from a convenience to a powerhouse. The term "desktop replacement" is now used in laptop advertisements; it's become a reality. If you are computer recording, the laptop is a very attractive choice because you can take it anywhere. The speed and disk size of current machines make them more than adequate for tackling even complex recordings. Pick up any recording magazine and you'll find some article explaining how your favorite artist's record was recorded live on a laptop at a gig. With the choices in USB and Firewire audio interfaces dropping in price and upping in features, laptops are a great choice.

The laptops themselves can be much more expensive than a desktop, so that is a major drawback. Laptops are also very hard to upgrade, and the jury is out on how durable they are in the long run.

Standalone Units

Standalone recorders such as ADATs, DA-88s, and hard-disk systems by Alesis and Mackie can be thrown in a rack carrier and taken from place to place. Throw in the necessary mixer, possible additional microphone pre-amps, or other extras, and the weight starts to add up. But it *is* possible to get very high-quality results from these standalone recorders. Studios-in-a-box can easily be taken around and are good portable solutions.

Field-Based Recording

If you are trying to record live sounds like the sound of the ocean, kids playing at a park, or a noisy city street for some added effects to your recordings, you can do so easily with either a compact DAT machine or a minidisk recorder and one stereo microphone. The quality will amaze you. This same setup can be great for picking up an acoustic orchestra or band that wouldn't require multitracking; most orchestral recording is done with one set of stereo microphones in front of the group. You can even record a live gig this way.

Setting Up a Space

Having a comfortable space to work is critical to working efficiently. If your gear isn't readily available to you, you aren't going to be as likely to use it, so don't cram yourself into a corner someplace or exile yourself to a basement.

Being comfortable is vital to working efficiently in any area to have things within hand's reach. You can either make do with the tables and chairs you already own or invest in studio furniture. Yes, they make furniture just for this kind of thing! But before you furnish your studio, you have to establish what your main focus is. For most people, easy access to their main instrument is their greatest concern. Then comes placement of the recording device or mixer. If you use a computer, do yourself a favor and put the computer on the floor if you can. This will free up much needed desk space. **FIGURE 3-1** shows a well-organized recording space.

FIGURE 3-1

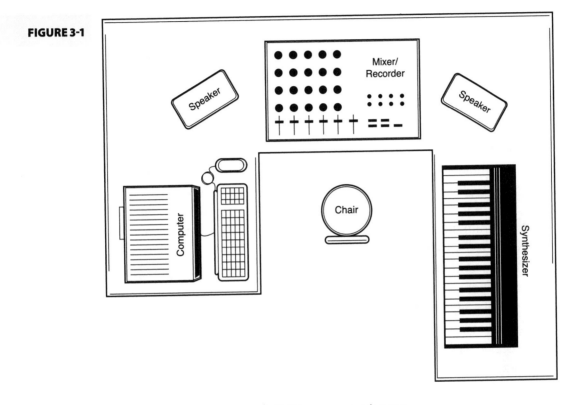

▲ Setting up a good space

You need a place where you can get work done. Your music requires concentration. Selecting a spot for your studio might not always be in your control, but you should take a few things into account when setting up your space. To record with microphones you will need someplace quiet. Microphones have this pesky little way of hearing things you don't want them to hear: dogs barking, doors shutting, and phones ringing, just to name a few. If you plan to work late at night you'll want to be someplace out of the way so you don't disturb anyone.

Instruments

With all this talk about equipment, you can't forget about the instruments you're going to play. They're pretty important to this process because, no matter how good your recording gear is, your instruments need to sound good. It stands to reason that any problems you have with your instruments are going to get worse when you record them. There's a widely used phrase in the recording world: garbage in, garbage out. Make sure you have decent-sounding equipment. It's a myth that "you can fix it in the mix."

Adding new sounds to your music is one of the fun parts about home recording. Even if you only play guitar, adding several different-sounding guitars on different tracks can widen the variety of sounds you can make. Experimentation is the name of the game here; you'd be amazed at what sounds good together. Keyboard players can really go to town with different sounds, layers of instruments, even drum kits from the keyboard.

Drum machines have always been useful to nondrummers and home studio musicians alike. Acoustic drums can be difficult to record well, and can be too loud for many apartments and houses. Drum machines not only keep time, but also can blend into the mix of sounds well enough not to stick out as "fake." Drum machines have also gone into the virtual world of the computer. Sample-based drum programs such as Battery from Native Instrument sound so realistic it's uncanny. They sound so realistic because samples are actual recordings of drums, not synthetic versions. Premade drum loops are every bit as real as having the drummer with you. These loops are professionally recorded in studios, are well mixed, and sound very cool—something worth checking into. To utilize loops, you would use a computer and recording software. Loops come as premixed audio files. You simply add them into your recording program on an empty track and voilà, instant drums! Loops are widely used in dance, hip-hop, and electronica.

Just because you *can* lay down eight or more tracks of yourself, doesn't mean that it's the best idea to do so. Why use a drum machine when a real drummer is close by? By getting to know other musicians and home studio owners, you can collaborate with each other and utilize the combined power of all your talents. Maybe you'll even make a record together? Who knows what could happen.

Chapter 4

Recording Equipment

The number of options available to the home recording market is staggering and, for many, a bit scary. Finding the right gear for you can be a tall order. This chapter covers noncomputer setups from the simplest to the most extravagant. One thing you will not find are specific recommendations on brands and models. Instead, you'll learn about all the options available to you, so that you'll have the information you need to decide what will work best for you.

What the Right Gear Can Do for You

There are many pieces of gear available, some fancy and some basic. Finding the right gear for you can mean the difference between creating music and creating frustration.

It's only in recent years that the sound quality and low prices of home recording equipment have made recording professional-sounding music at home possible. The early tape-based studios lacked the fidelity that professional studios could produce. With tape, it was easy to tell a homespun demo from a demo recorded in a professional studio. Digital gear has blurred that line substantially. The noisy recordings and tape hiss that plagued home studios are now a thing of the past. Twenty years ago, even an experienced engineer couldn't make a budget home studio sound truly professional; now, anything is possible.

It's not true that good recording equipment can make musicians sound better. Great gear can't improve musicianship, songwriting, or your general skill on an instrument. While signal processing can enhance your sound, the quality of the final product will primarily depend on two factors, no matter how much you spend on equipment. The first, and most important, is the quality of the sound you are recording. Unrehearsed, badly organized music and music played on poor-sounding instruments will sound lousy every time, no matter how expensive the recording gear is. The second factor is the knowledge and skill of the recording engineer. The art of microphone placements, EQ, and all the effects settings contribute to the quality of the sound. You get out what you put in.

Just as buying a Ferrari won't necessarily make you a better driver, buying top-level equipment won't instantly make your music sound better. The opposite is usually true; the better your gear, the more accurately it records and plays back, including both the good *and* the not so good.

It's time to break down gear into categories based on price, which is the determining factor for many people. Remember that prices might change over time, but the ideas behind what you get for the money are the same.

Every piece of home recording gear is trying to imitate the features of a professional recording studio on a smaller scale. While technological advancements might allow more features, the goal is still the same: achieve at home the quality that the professionals get in a studio.

The 4-Track Tape Recorder

More than twenty years after its initial introduction, the 4-track cassette recorder (shown in **FIGURE 4-1**) refuses to go away without a fight. Many home studio owners got their start with machines like these. Nowadays, this is one of the least expensive ways to get into home recording. If you are just getting into this, and money is at a premium, you can do very well with a tape multitrack.

FIGURE 4-1

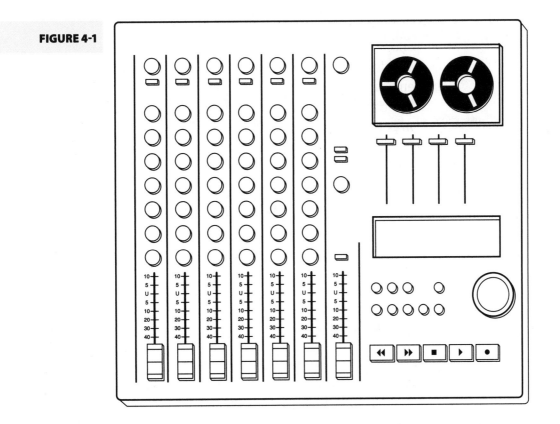

▲ 4-track cassette recorder

The cheapest multitracks go for about $100. These basic machines, currently manufactured by Fostex and Tascam, allow one track at a time to be recorded. However, a maximum of four separate tracks can be played back at once. You can control each track's level (volume) and the pan (left to right balance), but that's all the control you get. After you find a blend you like, you can output the recording to another 2-track "normal" cassette deck to capture the final mix. As for the quality, don't expect miracles. Even so, these recorders work well for documenting ideas and rough sketches, plus they are very portable.

As you step up in price, the number of features increases. For around $150 you can record two inputs at once. You still only get four tracks to work with, which might be plenty of tracks for you. At this price level there is no EQ or sound manipulation other than volume and pan.

For around $250 the feature set really spikes. Four inputs can be used at once and the units add high and low EQ. You also get the option of using effects plugged into auxiliary channels.

ALERT!

Don't know what an auxiliary channel is? Take a peak at Chapter 10 dealing with mixers to learn how to utilize auxiliary channels.

The best 4-track cassettes will cost you about $350. Eight inputs can be used at once and all four tracks can be recorded at the same time. High, middle, and low EQ are included for better control of the sound. The standard volume fader and pan knobs are found along with the auxiliary inputs for effects. You can make surprisingly good sounding demos with this machine.

Transition to Digital

Not long ago the cassette was the format of choice for home recording. But at the time of this writing, there are only five models left in production and only two companies continue to make them. It's likely the number of 4-track recorders available will continue to diminish because digital technology is available for the same price, and you get higher sound quality and added features. The end might be near for the cassette tape now that digital is here to stay.

Low-End Solutions

When you enter the digital recording market you instantly gain some nifty features. Built-in effects such as reverb, delay, and even guitar amplifier simulators are standard. The low-end digital recorders now record on compact flash or smart media cards. Compact flash and smart media are the memory modules originally used in digital cameras. They have now found their way into the home recording market.

For between $200 to $400 you can purchase a digital recorder. Zoom, Korg, Tascam, Boss, and Fostex are currently producing these types of machines. These low-end digital recorders let you play back anywhere from three to eight tracks at once. The built-in effects are a great addition. (But they won't sound as good as external effects processors.) The number of simultaneous inputs is usually small in this range, so don't expect to record more than two sources at once. If you plan to record track by track, all alone, that won't pose a problem for you.

One really neat feature that's showing up in these units is background drum and bass rhythm tracks. You tell the unit what style you want and how fast to play, and it creates the background music for you. And many of these units can run on battery power, making them great for taking with you to capture spur-of-the-moment ideas. Another feature that's showing up is an included USB cable for easy connectivity to a computer. You can use it to transfer the music to your computer and then burn a CD (if you have a burner).

FACT

With digital technology, the size of the storage or memory media determines how long your recordings can be. The more megabytes it has, the more music you'll be able to fit.

Of course, there are always some limitations. Some of the limitations of low-end digital recorders include:

- Fewer inputs than a similarly priced cassette studio
- Compact flash and smart media are more expensive than tapes
- Fewer knobs to turn, giving less control
- Editing on a small display screen

Mid-Range Solutions—Studio-in-a-Box

The next step on the ladder will take you up in price. Every jump in price means you gain something over the previous level, usually more inputs, better quality effects, and more support for multiple channel recordings. Hard-drive storage begins at around the $550 mark for our purposes and will top off at $1,000. At this level you start gaining more control over your sounds. The number of tracks you can play back is at least eight, and some models go higher, as high as sixteen. You also get into editing features at this level: the ability to move music around, cut and paste, and easily rearrange tracks. On many units you also find a built in CD burner. When you have finished your sessions, you can master to a CD. Small LCD screens are standard for accessing effect settings and editing the tracks.

Mid-priced hard-disk recorders (shown in **FIGURE 4-2**) can be purchased from Zoom, Tascam, Fostex, Korg, and Yamaha. Pay particular attention to the size of the hard drives. The bigger the disk, the more music you can store. Since the hard disks are buried inside the unit, once it's full, you have to finish the process and get the music off in order to record more. Compare models to see what's available. You can achieve very high quality results in this price range.

Top-of-the-Line Solutions

Units over $1,000 can be considered "top-of-the-line." Prices can shoot as high as $4,000. This category consists of studios-in-a-box and standalone recorders.

What do you get in a top-of-the-line studio-in-a-box? More inputs, higher quality, larger hard drives, bigger LCD screens for editing, more tracks, and other fun toys. As you climb the price ladder you get extra toys, including motorized faders, external computer displays, mouse inputs for editing, and digital outputs for mastering to DAT.

FIGURE 4-2

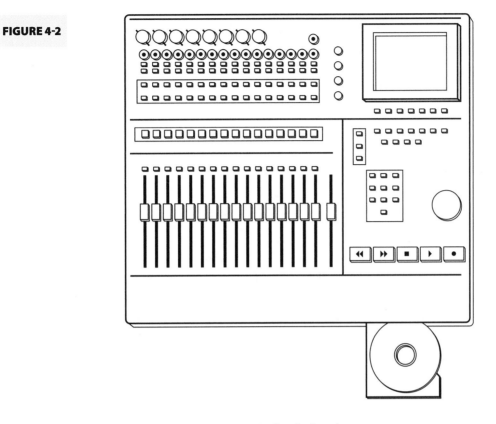

 Studio-in-a-box

Motorized faders let you record a mix as you go along. When you go to another part of the song, the fader remembers where the volume was at that point in the song and moves itself there. You can also "automate" a mix by recording the fader movements as you mix; they will play back by themselves.

Studios-in-a-box are serious systems, worthy of the name *workstations*. The quality of the internal effects, flexibility of editing, quantity of inputs, and support for more live tracks make these workstations "professional" quality. You can find high-end devices made by Fostex, Yamaha, Roland, and Akai.

Standalone digital recorders are units that only record audio. No mixing, no preamps, no effects, these units only record multitrack audio with professional quality. Why would you use these instead of an all-in-one studio? If you already own lots of outboard gear, tons of rack effects, and a mixer, then this might be for you. These systems are not for the first-time user! You will find these exact same units in professional studios all over the world. Popular units are made by Tascam, Alesis, Fostex, Akai, and Mackie. The quality of the recordings made on these is extremely high.

Although top-of-the-line solutions are beyond the scope of most home recording studios, it's good to know they exist!

Microphones

Microphones are covered extensively in Chapter 9, but for now let's take a look at some of the basics. When looking for a recording device, especially a studio-in-a-box solution, you need to figure out how many simultaneous microphones you want to use. Microphones don't put out enough energy to be adequately heard by themselves; they require microphone preamps to boost the signal. Recorders tend to skimp on the number of supplied microphone preamps, so check the specifications carefully to make sure that you have what you need.

If you find a great recorder at a great price with only two inputs for microphones and you would like to record four microphones at once, you're not stuck. You have the option to buy external microphone preamps that can plug directly into the line inputs; most recording devices provide plenty of line inputs. Buying more microphone preamps means spending more money, so do your homework when choosing a recorder. The microphones themselves have a wide range of prices, from $50 to as high as several thousand dollars.

Headphones or Speakers?

When it comes to listening to your recordings, there are only two options: headphones or speakers (called monitors). Headphones are the least expensive solution, and many musicians prefer to mix on headphones. When selecting headphones (shown in **FIGURE 4-3**), make sure to buy headphones designed for audio mixing. Audio mixing headphones should have a flat response: They don't boost the bass or lower the treble like other "radio" headphones sometimes do. Headphones start around $20 and top off around $150 or more.

FIGURE 4-3

◀ Studio headphones

If you want to listen without headphones, or you need to have several people hear the mix at once, you will want to go with stereo monitor speakers. Monitors resemble the speakers found on your stereo system. Monitor speakers should have a flat response like headphones, so you can hear what's really going on with your music.

QUESTION?

Why is "flat response" so important when mixing?

If the speakers unnaturally boost the bass, or reproduce any part of the signal unfaithfully, you might decide to change the sound to make it balanced to your ears on those speakers. After you finish and mix down to a tape or CD and play it back on a good sound system, you might find that the bass is too low because your monitor speakers emphasized the bass too much and you cut it during the mix to balance out. It's important to hear what's really there.

FIGURE 4-4

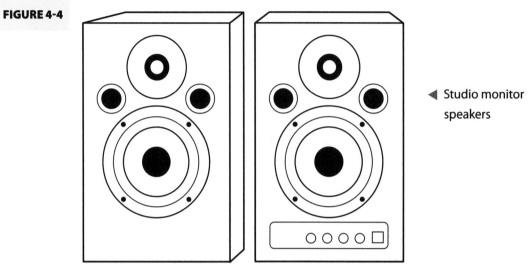

◀ Studio monitor
speakers

Two monitors (left and right) are the standard way to mix, as you can see in **FIGURE 4-4**. There are two types of monitors, self-powered (active) and passive. Just like a microphone signal, the output from the recorder won't be loud enough to drive the monitor speakers: it needs to be amplified. Active monitors are slightly more expensive because they contain amplifiers built into the speakers. Passive monitors require a separate power amplifier between the recorder and the speakers. In a home studio, active monitors are usually easier to deal with. You can expect to spend $100 on the low end to $800 or more on the high end.

Accessories

It's the little things that can eat up your budget. You'll need cables for every input you plan to plug in, and there are different cables for every type of application. Chapter 8 covers in detail the different types of cables and what they are used for. You might need microphone preamps or direct boxes for your setup. If you plan to burn a lot of CDs, include blank CDs in your budget. Adapters that take one type of cable and convert it to another can be very expensive. Have some extra cash left over for emergencies. If you have ever done a do-it-yourself home repair for the first time, you remember that you ran to the hardware store often. When you get started recording, you'll find yourself running to the music store often as well!

Sample Studio Setups

The studio shown in **FIGURE 4-5** is perfect for the beginning home recorder. A basic 4-track or small digital recorder is the center of this studio. This studio is designed for the person who is recording one or two tracks at a time, so having only a few inputs isn't a problem. For microphones, one simple dynamic and one condenser microphone are plenty (see Chapter 9 for more on types of microphones). The dynamic microphone is great for vocals and amplifiers, while the condenser handles everything else. The digital recorder has built-in effects; while the tape-based 4-track requires effects to be provided by external devices—either pedal-controlled floor units or rack-mounted ones. Monitoring and mixing are done through headphones or a stereo system. This is a great starter studio that won't cost an arm and a leg.

FIGURE 4-5

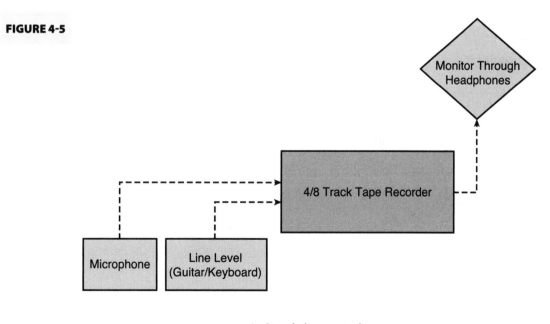

▲ Simple home studio

FIGURE 4-6 ▼ Standalone Studios

Studio-in-a-Box Standalone

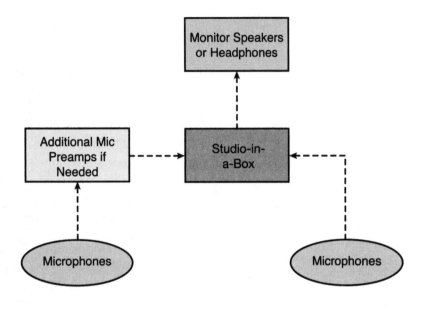

Standalone Digital

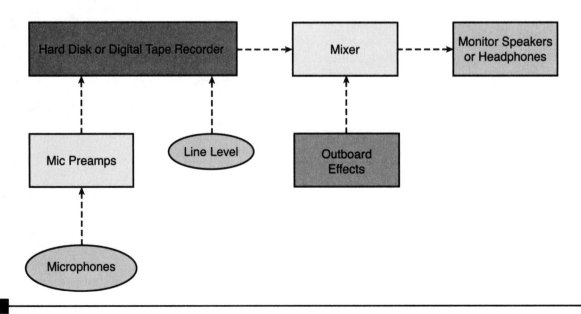

Those who opt for a standalone studio (as shown in **FIGURE 4-6**) will use either a studio-in-a-box or true standalone units like ADAT, DA88, or any hard-disk recorder.

Those opting for a studio-in-a-box won't need a ton of extra hardware, because all the major functions are self-contained. Internal effects, faders for mixing, CD burners, and multiple inputs are what make studios-in-a-box so convenient and popular. The number of microphones will depend on how many sources need to be recorded simultaneously. Studios-in-a-box, especially the lower-end models, don't contain many microphone inputs. This setup might require an external mixer or separate microphone preamplifiers to accommodate the extra microphones.

True standalone units like the ADAT or a hard-disk recorder require external mixers and outboard effects. The added cost of these extras make standalone units less desirable for home studios on a budget than studios-in-a-box. Monitoring for standalones is done through headphones or, preferably, through monitor speakers. The standalone studio shown here is a nice setup that is flexible and affordable for many.

Chapter 5

Recording on a Computer

In the 1990s, the personal computer began to enter the average home. Now, the computer is as standard as a couch and a TV. It's no surprise that the computer has been a valuable aid to music. In the past few years, the home studio owner has been reaping the rewards of the love affair between technology and music. Luckily, computer prices have fallen so low that powerful machines are much more affordable than in years past.

Where Did It Start?

For the home studio user, computer music history started with the invention of musical instrument digital interface (MIDI). MIDI is a standard language to allow electronic instruments and computers to communicate. The invention of MIDI led to the computer's ability to control a keyboard synthesizer. Unlike audio, MIDI does not have to be played in real time; it's a text file of commands, not sounds. Because it's just simple commands and not actual recorded audio, you can play MIDI parts one note at a time, as slowly as you want. MIDI is a form of electronic composition; you write it one note at a time and the computers/instruments play it back for you. Since MIDI issues simple note-on/note-off commands to control the keyboard, editing and manipulating MIDI music is very simple. The sequencer was born from this marriage of MIDI and computers. Sequencers can either be physical machines or computer programs. Nowadays, it's more common to use sequencing programs in your computer.

Sequencing

A sequencer functions much like a multitrack audio recorder. Tracks are recorded one on top of another and arrangements are built up one layer at a time. Since the sequencer doesn't actually make any music—all it does is control the keyboard, much like a player piano—the sequences can be highly edited. Just like book publishers reveled in the idea of being able to cut, copy, and paste text in a word processor, the creation of sequencers gave MIDI-based musicians the same power. Whole sections of music could be rearranged with ease, and editing could be as precise as note-by-note changes. Since sequencing didn't require a powerful machine to operate, computers of the 1980s could handle the job of sequencing MIDI. A great deal of the commercial music of the last twenty years has been a combination of sequenced and live music.

FACT

Sequencers continue to be a vital part of professional and home studios. For more about sequencers, check out Chapter 7 for the details on what programs are available and what they do.

Digital Music

The unbelievable editing power that sequencers afforded composers and musicians contributed to the growing need for the ability to edit audio as easily. Editing audio with analog tape meant cutting and gluing tape together on a splicing block. This was a very difficult and arduous task to do without the music sounding like it had been hacked up. Initially, the computer was used for stereo master mixes only. It was possible to do edits at high resolution on the computer screens. However, the computers had a hard time dealing with the large file sizes of audio and handling the complex processing needed to work with audio data. In time, multitrack computer audio became available and there is now an industry standard for multitrack audio: Digidesign Pro Tools. All of this technology came with a price, a price that was out of reach for almost all home studio owners. Recently, the power of the modern personal computer with its lower-cost, well-crafted software has allowed home musicians to join the party.

What It Can Do for Your Music

The computer has changed the way music is made. The flexibility of the current software and sound quality has made the computer an indispensable tool. You might be asking, "This is all great, but what can it do for me?" Here is a short list of what a computer can help you accomplish:

- Integrate multitrack audio and MIDI
- Edit and move music around much like a word processor lets you do with words
- Easily burn to CD and distribute your music online

In short, the computer can be whatever you want it to be. It can easily function as a recorder, sequencer, effects processor . . . you name it, a modern computer can handle the job. Now that you're convinced you want to go with the computer, you'll need to get the computer set up.

Processors and RAM

No matter how well your computer runs now, you'll need to tweak the setup for home studio use. Audio recording places special demands on the computer that are very different from the demands of surfing the Web or checking your mail. The more power the better here.

Starting Fresh

In order to really get your machine running smoothly, the best thing you can do is start over. If it's at all possible, and you know how to do it, back up your important data and reload a fresh copy of your operating system. While this might seem drastic, the majority of computers get lethargic and become error prone due to old files hanging around the system. Generally speaking, spring cleaning like this will always help. Starting fresh can breathe new life into a machine that starts to feel old and slow.

Consider dedicating one computer to music alone. No e-mail, no instant messenger, no Internet—only music. In professional studios, a dedicated audio computer is standard. Think what would happen if you got a virus on a computer that holds all your music. You could lose everything!

Processor Speed

The processor, properly called the central processing unit or CPU, is the brain of the computer. The speed of the processor is the first consideration. It's all about speed! The speed of your computer is measured by the frequency at which the processor is able to perform an instruction, called instructions per second (IPS). This number used to be stated in megahertz (MHz), which implied a million instructions per second or MIPS. Now that chips are faster than 999 MHz, the term gigahertz (GHz) is used for any chip that exceeds 1,000 MHz (1 GHz equals 1,024 MHz). What do megahertz and gigahertz mean to you? The higher the number, the better, because the higher the number, the more "tasks" the computer can do at once and the faster it can do them. More powerful processors are able to play more tracks, add more effects, and perform more elaborate edits, and so on. Unlike studios-in-a-box and analog tape machines, computers come with no guarantees on how many tracks and effects you can run in the software. Your possibilities (or limitations) of what you can do with software are largely based on how fast your machine is. But many variables—not *just* the CPU—affect what the computer can do.

Different Types of Processors

When you compare machines from the IBM world (PCs) that run the Microsoft Windows operating system with the Apple/Mac world, you might notice that the Apple/Mac processors have a lower stated speed when compared with the PCs. On the surface, this seems to indicate a decreased processing power, but this is not the case. Apple/Mac's processors are very different from the Intel/AMD processors used in PCs, and they cannot be compared simply by their processing speed. What's important is to look at the relative speeds of other machines. Compare the speed of one Mac to other Mac machines currently in production. Compare PCs to other machines that run Microsoft Windows.

RAM

Random access memory, or RAM, is another vital system component of your computer, also another number you want to be high. RAM is a specialized area where data is stored temporarily while the computer is on. It is called volatile memory because it is gone when you turn off the computer. RAM is superfast and data can be written into it and read from it at much higher speeds than from the hard drive. RAM comes in the form of an integrated circuit, sometimes called a chip, which connects directly to your motherboard. Installing RAM is not difficult, and the price of RAM chips has fallen dramatically.

Computers place into RAM memory important information that needs to be accessed quickly. RAM is measured by how much data it can store at one time. Having a lot of RAM will speed every computer up, no matter what speed your processor is. You could have the fastest processor on the market, but with only a small amount of RAM, the computer will crawl. Most software manufacturers suggest minimum and recommended RAM amounts. Check out Web sites and call companies to see what they recommend. To run music-recording software, most computers need more RAM than what they typically come with. However, computers have limits on how many RAM chips can be placed in one machine, so your ability to increase RAM is not infinite. Pay attention to this when you are purchasing equipment.

FACT

Generally speaking, you should fill your computer with at least half of the total RAM it can hold. If your machine can hold 1,024 megabytes of RAM, 512 megabytes of RAM will set you up well.

Your magic number for RAM will differ based on what you do with the computer. If you plan on recording only a few tracks, and not going crazy with effects, filling your computer with RAM won't be necessary. However, if you plan to use high track counts (sixteen or more), or do anything with samplers or virtual instruments(see Chapter 17), RAM is crucial. In this case, you can't have too much.

Hard Drives

There are a few attributes of hard drives to consider. The first and most obvious is the size. The larger the disk, the more music you'll be able to store. Like RAM, this number is expressed in megabytes or gigabytes. How big should your drive be? Buy as much hard drive as you can afford. More is definitely better in this case.

The next critical factor in a hard disk is the rotation speed. A hard disk spins around in the same way a CD does. The speed at which it spins is measured in revolutions per minute (RPM). The higher the speed, the faster the disk can access its data. Why is this important to you? Faster disks equal higher track counts in the software you use. If you have a fast CPU with tons of RAM, a slow hard drive will still limit you. The faster the better. You will also find recommendations for hard-drive speed listed on audio software manufacturers' lists of recommended hardware.

The last factor is seek time. Seek time is how fast the data on the disk can be accessed. Seek time is measured in milliseconds. The lower the seek time, the better.

I'll Take Two

Chances are your computer came with only one hard drive, and this works fine. You can record and store files on a single hard drive. However, the best way to go is to have a second hard drive dedicated to audio. Why is this? Simply put, if you have one hard disk, the computer has to use the disk for running the operating system, running any open programs, and recording huge music files. This is a bit much to ask of just one disk. Your track count will always suffer by using one disk.

Digital audio requires a good deal of storage space on your hard disk. Every minute of every track you record takes up many megabytes of space. As the quality of the recorded audio improves, file size increases as well. This is why large hard drives are crucial.

You don't have to run out and buy a second hard drive right away, however. It's best to start out with one drive and see if you overtax the machine. If you plan to record low track counts, this isn't much of an issue and you might never get an error. Users who push the computer with high track counts will get errors because the computer can't stream data fast enough to keep up. If you reach that point, get another drive just for audio.

Internal or External?

If you opt for the second drive, you have a choice: internal or external. Some computers won't accept a second internal drive, so that choice is made for you. Neither one is preferable over the other; both get the job done. The only advantage for an external drive is that you can take it from computer to computer. If you collaborate with other home studio users, this could be a big plus for you.

The Great Debate

Here we go . . . the big topic that has been debated and argued about for years. Should you use an Apple/Mac or Microsoft Windows? Each side has its strong and weak points. Both will let you run a studio. Let's consider each side separately.

Apple/Mac

The Apple Mac has one thing going for it over Microsoft Windows. Historically, Macs were the first computers to run music software, and so more software was written to run on the Mac. Most professional studios still rely solely on Macs for audio. But now, Windows has caught up with the Mac for music. Even so, the Mac has a particular working style that appeals to some. You have to spend some time using the Mac OS to see how different it is from Windows.

The only company that makes the Mac is Apple. Apple is the only game in town, and that's both good and bad. On the plus side, there is very little variation in the hardware, so software companies have an easy time making products that are compatible with Macs. On the downside, you have fewer choices for machines. There are some software differences that might sway your decision. Some software is still Mac only, with no Windows version. You have to do some research. Macs tend to be slightly more expensive than Windows computers, but that's becoming less of an issue as prices equalize.

Microsoft Windows

The majority of the desktop market is PC-based, and most of those computers run Microsoft Windows. Windows is now very capable of running music software as well as the Macs do. Lots of music software is now available for Windows. With Windows, the version of the operating system is critical for music applications. Many music applications will run on only the latest version of Windows. Be careful to check that you have a machine capable of running current software. Makers of software for Windows have the unique challenge of trying to be compatible with literally millions of different hardware combinations. Unlike Apple, many different companies make Windows-based PCs. Apple makes all its hardware, which is designed to run its exclusive operating system, whereas Microsoft Windows is a program run by computers made by a variety of different companies. Each computer running Windows uses different CPUs, different RAM . . . you get the idea. Sometimes software compatibility can be a problem on the Windows side because of this.

All your music applications will take up a lot of screen space. Having a large monitor will make your life much easier. Being able to have several windows open at once will save you time. Prices of monitors have come way down with the introduction of flat-panel LCD monitors. You can pick up a very large monitor inexpensively. LCD monitors are also really nice, but also more expensive than the old-style CRT tube monitors.

In the end, the choice is yours. Whatever you have, you'll be able to run some kind of music software. If you're thinking of buying a second computer just for audio, make sure to give both sides a fair look.

Interfaces

So now that we've covered what types of computer hardware make a good audio system, it's time to turn to how to get music in and out of the computer. It's time to talk about interfaces, those wonderful devices that connect music to an otherwise lifeless machine. Interfaces are pieces of hardware that connect to a computer to bring music in and out.

MIDI

MIDI interfaces are the simplest and least expensive interfaces for a computer. They come in many shapes and sizes, one for every need. You will want to get an interface that has one input for every piece of MIDI-enabled gear in your studio. The MIDI interface is shown in **FIGURE 5-1**.

FIGURE 5-1

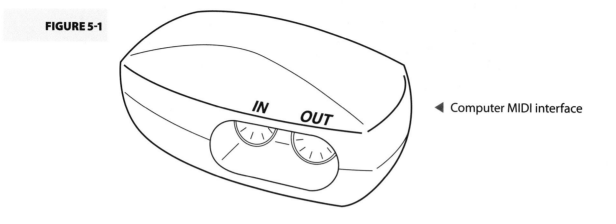

◀ Computer MIDI interface

Basic interfaces start around $35; an interface with more inputs will be more expensive. In terms of connections, MIDI interfaces come in a few flavors:

- **Serial:** Not very common now. This is a connection that attaches to a serial port on your computer.
- **PCI MIDI interface:** A card that sits inside the computer. Usually combines audio and MIDI.
- **USB MIDI interface:** A small rectangular port on the back or front of your computer.
- **Firewire MIDI interface:** Firewire, as it's known on the Mac, or IEEE 1394 as it's known in the Windows world, is a new connection that is becoming standard. Firewire interfaces usually combine audio and MIDI.

You will need a MIDI interface if you plan to record MIDI from keyboards, synthesizers, or drum machines. If you plan to use the computer as an audio recorder only, a MIDI interface is not necessary.

Getting Audio In and Out

In terms of routing audio in and out of your computer, you have to make some hard decisions about how many instruments you can record at once.

Simple interfaces that support one or two channels are relatively inexpensive. If you are looking to record eight simultaneous inputs, be prepared to pay more. Also key is the number of microphone inputs that the interfaces have. If you plan to record acoustic instruments such as piano, voice, or anything else that requires a microphone, you'll need a few microphone inputs (shown in **FIGURE 5-2**).

FIGURE 5-2

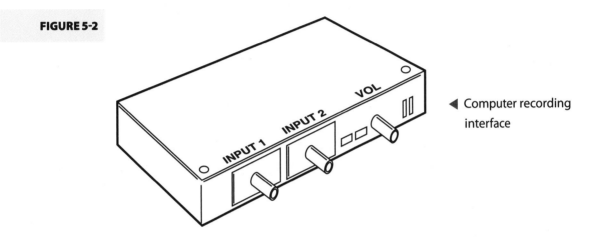

◀ Computer recording interface

Just like MIDI interfaces, audio interfaces come in three basic forms:

- **PCI:** A card that sits inside your computer. A cable usually attaches from the card to a tabletop unit, called a breakout box, that contains the actual inputs. These are the most common audio interfaces.
- **USB:** USB makes it easy to move between computers, but sometimes it lacks the necessary speed to pipe audio in and out fast enough. This is why USB interfaces rarely exceed four inputs. If you're looking to record many sources, USB might not be for you. They do work well for smaller recording needs, and they are priced attractively. If you have a laptop, you don't have the PCI card option, so USB is a choice for you.
- **Firewire:** Firewire is much faster than USB and allows for a greater number of inputs. It's also more expensive than USB. Firewire is becoming a great option for mobile and home recording because of its high track counts and portability factor.

Making It All Work

There are several things you can do to keep the audio computer running smoothly. First is unplugging the Internet while you work. If you dial in, or use broadband, the computer is always doing something in the background that is Internet related. This can take away from power needed in your music program. Log off, or power down your DSL/cable modem. On the same theme, don't run other programs in the background while you are running music applications. Music applications ask a lot of the computer, so give it as much brain power as possible.

QUESTION?

How will I know what to use when there are so many choices?
Ask people, ask music store employees, ask other home studio owners. You can even read the various Internet bulletin boards devoted to computer music for information on what works and what doesn't.

Without getting too technical, a hard drive can write data in many disparate places on its surface. Because the files aren't in one line, it takes longer to read them. This can be a problem. Defragmenting your disk (defrag) helps put the files in order for faster access, giving your disk more power. Power equals more tracks and more effects. Defrag often!

There are also many utilities to ensure that your disk is healthy and free of corrupt, evil files (known as viruses). You should also run these often to keep your disk in tiptop shape.

Always back up your work! You'll feel very bad if your hard work magically disappears, or worse, you catch a virus that wipes out all your data. You can back up by burning data to a CD or a DVD. Both are cheap and very reliable. DVDs hold 4.7 gigabytes of information, while CDs hold 700 megabytes. There are other ways to back up, using tape drives and such, but CD/DVD is the most common.

Computers are known for doing some strange things to your data. Back up! You'll be glad you did when disaster strikes.

Sample Studio Setups

The computer-based studio is scalable based on what you want to do. One of the best uses of computer studios is for keyboard players who use MIDI. Get yourself a simple MIDI interface and some sequencing software and you're good to go. No microphones are needed if all you do is MIDI.

FIGURE 5-3   ▼ Computer-based studio

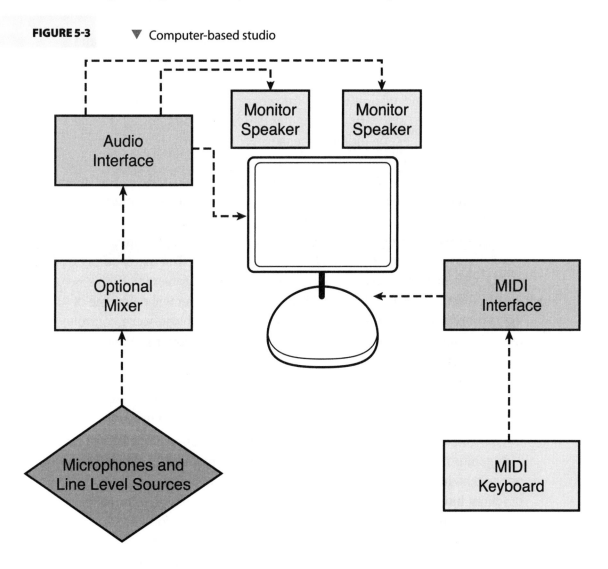

For audio recorders, the computer interface is the most crucial element here. How many inputs do you need at once, and how many have microphone preamplifiers built in? Interfaces start cheap for a few inputs and go up from there. Microphone channels are always key here. The computer acts as a digital audio workstation (DAW), handling the audio in the computer. All of the mixing and effects are applied in the computer. The final mix will be bounced into a file that can be burned on the computer's CD burner. The number of microphones will depend on how many and what kinds of sources you're going to record. With the right interface and a fast enough computer, it's possible to record eight or more channels at once. Playback and mixing can be anywhere from sixteen channels up to as many as your software or computer can handle. Monitoring is done through headphones, or monitor speakers. The majority of professional and home studios are now computer based.

FIGURE 5-4 shows a studio setup for recording several players live. As soon as you need the ability to record eight or more sources at once, you either have to make compromises or step up into the higher ranks. A compromise would be to use a large mixer to handle the multiple sources and mix down to a few outputs—usually two or four outputs. The mixer outputs can be sent to a computer or standalone studio. The problem with this is the inability to control all the sounds afterward. The mixer will pair some of the microphone and line sources together and you will lose some control when you mix. In this case, getting a good sound on the board is crucial to your success.

For those who need control of every source, you will need to purchase either a fairly extensive computer recording interface that handles eight or more channels, or one of the higher-end studios-in-a-box that supports the adequate number of microphone channels. Of course, you'll need a high number of microphone and microphone preamplifiers for each microphone that you connect. Many of the high-end computer and standalone products have multiple microphone input channels with preamplifiers. Monitoring becomes more complicated, because most performers want to hear the mix as they play. A headphone amplifier/distributor is necessary to power multiple headphones for all the players, not to mention all the headphones you'll need. Monitoring during recording is done with headphones and, preferably, during mixing, with monitor speakers.

This is one of the most demanding home studio applications and the price of getting this done will reflect that. However, if you are equipped to handle a live band, there's not much you can't do in your studio.

FIGURE 5-4   ▼ Studio for multiple live players

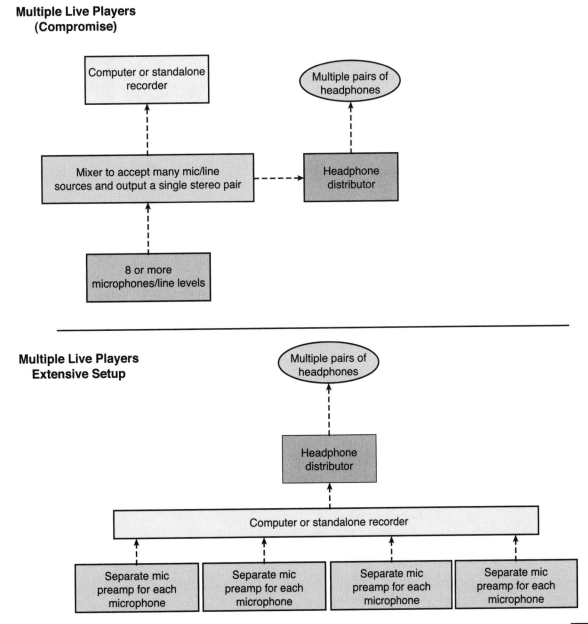

Multiple Live Players (Compromise)

Multiple Live Players Extensive Setup

Chapter 6

Computer Recording Tools

Twenty years ago the personal computer was in its infancy. The last place you'd expect to see one was in a recording studio, other than in the front office for billing and records. When computers finally made their way into the audio world, they were considered the dreams of tomorrow. That day has come, thanks to powerful computers and groundbreaking software. What we have now is nothing short of a revolution—a home studio revolution—thanks to technology.

Power in a Box

While we could pay homage to the computer and its miraculous power, if it weren't for the evolution and development of software in the audio field, computers would be nothing more than glorified calculators. The software available today re-creates the recording studio inside a personal computer. Mixing, editing, and effects can all be achieved seamlessly in a virtual environment, mimicking a real recording studio and its hardware.

Software studios have grown dramatically in the past few years and now make up the largest segment of the home studio market. What makes computers so great? For many users who already own a fast computer, the addition of recording software turns their ordinary home computer into a full-blown music studio. It also takes up little space, which is a big plus. Nowadays, the computer is becoming the central fixture of our life. Why not put our music studio there as well? Software manufacturers have answered the call with outrageous software for music production.

Advantages of Software-Based Systems

There are many good reasons to use a computer to run music software. First, today's music software is state of the art: It's where the field is headed and where all the cool advancements are showing up. Computers take up little space and can be upgraded slowly over time. Many people feel that editing with a mouse on a screen is the best way to go. Budgetwise, you might do much better with a software studio vs. a hardware studio like a studio-in-a-box. If you're technical in nature, you might find the working style of computer music suits you very well.

Disadvantages of Software-Based Systems

It's not all good news . . . If you don't already own a suitable computer, the expense of buying software and a new computer could prove to be much more expensive than the hardware alternatives. For anyone who's worked with computers, it's a known fact that computers are prone to crash, catch viruses, and eat files. And let's face it, some people are not tech savvy and just wish to "make music" without worrying about getting a faster hard drive or defragging their disk. What if you want to take your killer studio to a rehearsal and track your band? Going to haul the whole computer? No, not likely. There's also a group of people who swear that digital recordings never sound as good as analog ones. The debate continues on that topic.

The Interface to Software

Computers are highly adaptable machines. They can handle almost anything you throw at them . . . that is, if you can get your information into the machine. In the case of music, you need a special interface to get sound and/or MIDI into the computer. In addition to the software component, the interface is just as important, in many cases more important. Let's break down the interfaces that you will need in order to get sound and MIDI into your machine.

MIDI

Of all the interfaces, the MIDI interface is the simplest for the computer. MIDI format consists of simple data that can easily be streamed to a computer. You can get a simple MIDI interface inexpensively, and many companies make interfaces to work with every make and model of computer. The MIDI format is often included with the PCI, USB, and Firewire interfaces.

PCI

PCI (peripheral component interconnect) is a standard card that sits inside a desktop computer and adds audio functionality. PCI cards can range from simple one-input/one-output configurations to extensive audio options with many inputs and outputs. The more extensive inputs and outputs usually plug into a box called a breakout box that is connected to the PCI card via a special cable. Unfortunately, laptops can't use PCI cards without expensive modifications or expansion chassis. For desktop computers, PCI is a great way to go and there is a nice variety of these interfaces for every budget.

USB

USB (universal serial bus) is a relatively new way to connect peripherals such as mice and joysticks to computers. It was only a matter of time before audio interfaces started to appear using USB. One of the greatest benefits of USB is that it is compatible with desktop *and* laptop computers. USB is simply a kind of hardware port on computers that allows users to attach devices to the computer using a USB cable. They are now standard on all machines. USB devices simply plug into the back of the computer; no work inside the machine is involved. This makes portability a possibility with a computer studio. Economically, USB interfaces are reasonably priced. However, due to the way USB pushes data back and forth to the computer, don't expect a ton of inputs and outputs. USB is great for small setups and portable solutions. USB 2.0 is a newer form of USB using the same standard interface but transferring data at a much higher rate. Expect to see USB 2.0 interfaces capable of many simultaneous inputs and outputs soon.

Firewire

Firewire or IEEE1394, as it's also known, is a high-speed interface that, like USB, no longer requires a card inside a computer. Firewire is another hardware port on the computer that allows devices to plug directly into the computer using a Firewire cable. Firewire was originally conceived for digital video cameras to transfer large amounts of video data to a computer at high speeds. As an audio interface, Firewire is very popular because it can handle a large data flow and many simultaneous inputs and outputs. It's also ideal for laptop computers that need a powerful audio interface. Firewire is fast becoming the preferred audio interface offered today.

No matter what interface you choose, make sure it offers the connectivity you need. Look especially hard at the number of microphone inputs your interface allows.

The Curse of Latency

Computer systems are really great . . . but there is one catch: *latency*.

When you are plugged into an amplifier, you expect the sound to come out immediately after you play, right? You would expect the same from a recording system: You plug in and hear the signal in real time. But this is not necessarily the case with a computer. Computers deal with only digital information, so they have to convert your audio (analog) signal to digital. The computer then has to store it somewhere and retrieve it to send it back out. Then it must convert the digital signal to audio again. The problem is that the process takes some time, on the order of a few milliseconds. One millisecond won't feel like much, but approach ten or more and it starts to feel lethargic. This has been a problem from day one, but it's only getting better—that's the good news.

Computers are getting faster and can do all the conversions quicker. Interface manufacturers have also smartened up and added features that give low or no latency monitoring, which cuts down greatly or eliminates latency. There is usually a catch with low-latency modes; you usually lose the effects from the computer. So if you're recording a vocal part and you want to monitor the signal with reverb, you need a hardware reverb processor, you must sacrifice the reverb, or you must deal with the latency.

Don't be too worried about latency. Latency times have decreased significantly over the past few years, so now it's just a minor annoyance. At the speed technology moves, latency should be a nonissue soon.

Types of Music Software

Music software falls into a few categories. But the lines between the categories are blurring, even as we work. In the last ten or so years, three main types of computer music software have been developed.

Audio Multitrack Software

Audio multitrack software attempts to re-create a multitrack recording and playback studio inside your computer. When recording software was first introduced, massive PCI cards and other hardware were needed to help the computer cope with all of the audio data. A good example is the professional Pro Tools audio software that still, to this day, relies on PCI cards for computer power. Nowadays, software that deals exclusively with audio and not MIDI is very difficult to find, because most, if not all, studio software incorporates MIDI in some regard. But if you look hard enough, you might find some free or very cheap software on the Internet that deals with just audio.

MIDI Sequencing Applications

Before audio was even a dream on a computer, there was MIDI, which consists of small text commands to control the playing of synthesizers. Computers were able to work with MIDI data almost twenty years ago. A MIDI sequencer lets you record and manipulate many tracks of MIDI information, allowing the computer to play back long, complicated parts that might be unplayable by a single human. You can create piano parts that are faster than anyone can play. A single person can program an entire orchestra to play back their music. Editing and manipulation was possible on even the minutest of levels. MIDI was the reason that the computer made its first appearance in the recording studio. Even today, you can still get MIDI-only sequencing programs. Some of the more famous programs that were the pillars of the MIDI sequencing world, such as Logic and Cubase, have grown up to include sophisticated audio features as well. So just like audio-only applications, it's hard to find just MIDI; most are integrated.

Integrated MIDI and Audio

Today, this is where most of music software is going—the integration of MIDI with audio. As all of the programs grew up, the need for an all-in-one solution became necessary. The software suites that dealt strictly with audio, such as Pro Tools, eventually adopted MIDI. Conversely, the MIDI-only camp grew audio wings. All the programs covered in this chapter allow you to record and edit MIDI and audio together in the same program. With audio programs, not only can you play and record, but you can process effects and perform exacting editing, which is what makes these programs special. On the audio side, nonlinear editing is the distinguishing factor that makes the computer more than just an emulation of a multitrack recorder. What's nonlinear editing? Read on!

What Nonlinear Editing Means to You

Let's start with linear editing. Think of an audiotape. Suppose you were recording a mix for your car's tape player. Later on you decide that the track you placed first on the tape you'd rather have at the end of the tape. You would have to re-record the entire tape to rearrange the order of tracks. There is no way to just magically "move" that song. You can't do this because audio tape is linear—it's read in a line and whatever appears first will be played first and so on.

Computer audio systems don't rely on tape storage; instead they use hard drives to store data. Although we hear this data as music, the computer can't distinguish between math and music, and that works in our favor. Audio data is stored on the hard disk in a nonlinear manner, which allows you to change the order of tracks and move whole sections of your song with ease. This is one of the main reasons computer audio took off. Editing is far superior on a computer system. It's easy to imagine that audio on a computer is processed much like text in a word processor—you're free to cut, copy, and paste as you wish. Audio data is treated the same way, and that's nothing short of revolutionary!

Audio-Editing Applications

Historically the first audio recording on a computer was not multitrack audio. It was simply stereo files. These files were loaded into an audio-editing application that allowed nonlinear editing in high resolution. After the edits were completed, they were sent back to the tape they came from. This was the beginning. Even as multitrack grew up, audio-editing applications were still popular ways to edit in high detail. Nowadays, audio-editing programs like Wavelab are used for mastering and remastering because they don't deal with multitrack or mixing data, only the final stereo file. For many, audio-editing applications like these are the last step before burning or CD duplication. These programs are also handy if you work on a cassette multitrack and wish to mix down to the computer by recording the final stereo output into the computer. Some tape lovers even do editing this way!

Proprietary Audio Systems

By proprietary audio systems, we are referring to audio software and hardware packaged together under a brand name and sold for use on your computer. Pro Tools is an example of such a system and is, in fact, one of the only proprietary systems on the market today that can be used in the home studio.

Unlike Pro Tools, almost all audio software and hardware systems are modular, meaning you can buy software X and audio interface Y from different companies, and they will work together. For example, you don't have to use Digital Performer with a MOTU interface just because they are made by the same company. So too, you can get standard drivers that allow you to use almost any piece of software with almost any piece of hardware, regardless of the manufacturer.

FACT

Pro Tools is a proprietary system that doesn't work with other hardware or software. However, Pro Tools Free is a free downloadable version of Pro Tools that will work with any audio hardware, albeit limited to two inputs at once. Remember, *free!*

Pro Tools is not software, nor is it hardware. It's a proprietary system of hardware and software that is sold together for use on your home computer. Because they sell you everything at the same time, it's a self-contained system. This means you can't run Pro Tools software with any other hardware besides the hardware that comes with the package; but you *can* use Pro Tools hardware with other applications. This is not a bad thing if you love Pro Tools and wish to spend all your time working with it. Many people get confused by this and think they can plug their sound card into Pro Tools.

There are other proprietary systems available, but not for the home studio market. The only proprietary systems available in the home studio market are the LE line of Pro Tools, such as Mbox, and Digi 002.

Plug-In Formats

If you're working inside a digital audio workstation (DAW) and you want some nice reverb, or maybe a compressor or two, you're going to need a plug-in. A plug-in is the software equivalent of a hardware effects processor. Depending on what recording software you opt to run, you might need different types of plug-ins. Each recording program requires plug-ins to be written in a specific language that the host program will understand. Luckily, most plug-in manufacturers include multiple versions when you purchase, but not all. Let's break down all the major formats so you can see what's around.

VST

VST (virtual studio technology) is a plug-in standard format created by Steinberg, which also makes Cubase and Nuendo recording software. VST plug-ins work not only in Cubase, but other programs have adopted the use of VST plug-ins. VST plug-ins are available for both Macs and PCs. Because VST has been around for a long time, there is a nice selection of VST plug-ins available.

RTAS

RTAS (real-time audio suite) is the only plug-in format that works inside Pro Tools. Because of the widespread use of Pro Tools systems, there are many RTAS plug-ins available on the market. Because Pro Tools runs on PCs and Macs, you can find RTAS for both. The British company FXpansion has created a VST to RTAS converter allowing the substantial amounts of VST plug-ins to be utilized inside Pro Tools.

DirectX

DirectX is a Windows-only plug-in format that is used by Sonar and other popular Windows audio programs. DirectX was originally a multimedia language introduced by Microsoft to write games and other multimedia programs. Because of its integration into the Windows operating system, DirectX is used by many of the Windows audio programs. Sorry Mac people, Windows only!

Audio Unit

Audio Unit is a new type of plug-in format introduced by Apple as part of its remake of the Mac operating system version 10 (also known as OSX). The thought behind Audio Units was to create a system-level plug-in format that worked inside the operating system and could be available to any audio program on the system. Unfortunately, the only two major software makers using Audio Units are Logic and Digital Performer. Other programs use Audio Unit plug-ins on the Mac platform—Logic and Digital Performer are the major software platforms that do. Apples free recording software "Garage Band" utilizes Audio Unit plug-ins as well. This might change because the Audio Unit is in its infancy on the Mac platform. Audio Unit plug-ins are available only on the Mac platform in OSX or higher.

MAS

MAS (MOTU audio system) is the plug-in standard written by MOTU for use in its Digital Performer application. Before Version 4 of Digital Performer, MAS was the only plug-in that could be used in Digital Performer; however, now Digital Performer uses MAS and Audio Units. Since Digital Performer is Mac only, no Windows versions of MAS plug-ins exist.

Wrappers

A wrapper is a program that converts plug-ins from one type to another type. On the PC side there are DirectX-to-VST wrappers that extend the choices of Windows plug-ins. One the Mac side there is a VST-to-Audio Unit wrapper that greatly improved the number of available plug-ins while OSX was just starting out; however, developers have been slow to write Audio

Unit versions of their plug-ins. That is getting better over time as Mac OSX continues to evolve.

Popular Software

There are countless programs available for making music on a computer. However, a few have emerged as top players, and those are the ones we'll focus on here. All of these programs basically do the same thing—allow you to record MIDI and audio and to arrange and mix music, all on a computer system. The only differences are how they go about the task.

Digidesign: Pro Tools

No other name is as synonymous with studio recording as Pro Tools. Until a few years ago, the only system you could get was its expensive TDM system, which went for well over $10,000. In 1998, Digidesign introduced Pro Tools digi001 and entered the home studio market. Currently, Digidesign markets three home studio products—the Mbox, a two-channel USB interface; the digi002, an eighteen-channel Firewire interface plus motorized control surface, and the digi002 rack, an eighteen-input Firewire interface. All Pro Tool systems ship with the same version of Pro Tools LE software, which is capable of playing back thirty-two audio tracks and unlimited MIDI tracks. One of the biggest things Pro Tools has going for it is compatibility: Anything created on a Pro Tools home system can be taken to a larger professional version. This is great when you want to share your ideas or get expert mixing and mastering. As discussed earlier, Pro Tools is proprietary so you can run its software only with Digidesign hardware. As a music tool, it's a mature product that works on both Mac and PC equally well. It's also completely portable because the current versions work on USB and Firewire formats—great for laptop use.

It's set up simply with two windows—the mix and edit windows. The edit window shows you all your audio and MIDI data track by track, while the mix window shows the virtual mixing board and access to all your plug-in effects. A nice selection of audio effects is included with the package. You can always extend the system by adding RTAS plug-ins.

FIGURE 6-1 gives you a look at what Pro Tools looks like in action.

FIGURE 6-1   ▼ Digidesign Pro Tools   *Screenshot used by permission of Digidesign/Avid Technology, Inc.*

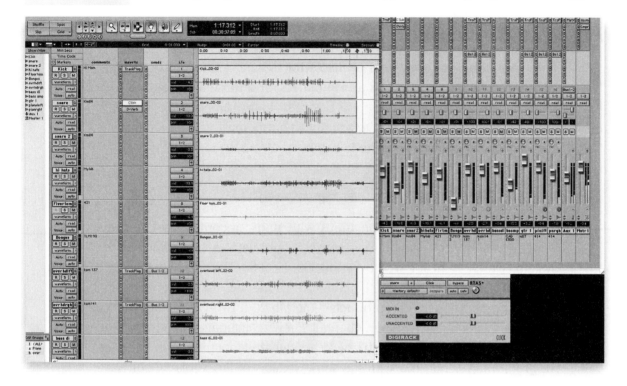

Steinberg: Cubase

In the 1990s, Steinberg introduced Cubase VST, an integrated virtual studio for music making. It has grown over the years and remains a very popular choice on both the PC and Mac platforms. It can run with any computer audio interface you choose. Like all programs of this type, MIDI and audio are grouped together. Cubase is a fully featured studio capable of anything you throw at it. There are a large number of plug-ins available for Cubase in the VST format. Steinberg also includes a nice set of VST audio effect plug-ins to get you started.

Cubase is also a very capable MIDI editor. This is because Cubase started its life as a MIDI sequencer and added audio capabilities later. Because of its clean format and ease of use, Cubase remains a very popular application. It is also completely cross-platform, running identically on both Mac and PC; a feat matched only by Pro Tools. Cubase comes in three flavors: SX, SL, and SE. Each version is more powerful than the next. Cubase SX is their top-of-the line application. **FIGURE 6-2** shows what the current version of Cubase, Cubase SX, looks like in action. Full automation is also included.

FIGURE 6-2   ▼ Steinberg Cubase SX   *Screenshot used by permission of Steinberg Media Technologies.*

Mark of the Unicorn (MOTU): Digital Performer

On the Mac platform (sorry PC users), MOTU Digital Performer is a favorite among musicians. Originally a MIDI sequencer that added audio capabilities, Digital Performer (or DP, as users call it) is another robust and powerful audio and MIDI tool. It has a clean interface with great audio effects and powerful MIDI editing. You can find Digital Performer in many composing and film-scoring studios around the world. The latest version of Digital Performer extends its power by adopting the Audio Units plug-in standard, greatly increasing the number of available plug-in effects. Of course, you get a nice starter set of audio effects from MOTU. Take a look at Digital Performer in action in **FIGURE 6-3**.

FIGURE 6-3   ▼ MOTU Digital Performer   *Screenshot used by permission of MOTU.*

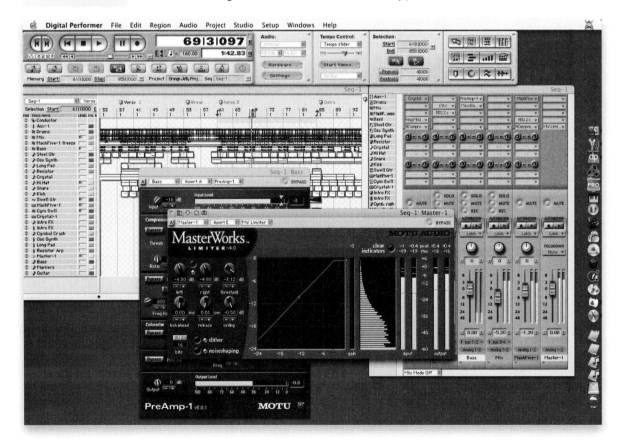

Emagic: Logic Pro

Logic, like many other programs, started its life as a MIDI sequencer and added audio later. Logic is currently produced only for the Mac platform, but it was at one time cross-platform. Logic is a very different program from the others covered so far. It is by far the most configurable and programmable software available for music making. It's almost a programming language wrapped in a music application. Don't let that put you off, though, because Logic is intuitive for basic MIDI and audio recording, and users who get into the underlying layers will find great power and flexibility. Logic sports many different windows and views for editing information. It boasts some of the best MIDI editing around. It also includes score editing to view your MIDI as music notation.

FIGURE 6-4   ▼ Emagic Logic Pro *Screenshot used by permission of Emagic Soft and Hardware GmbH/Apple Computer.*

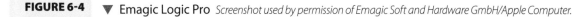

Logic is a very popular program that is starting to show up in more and more professional studios due to its powerful mix of audio and MIDI adaptability to any situation. It's also one of the most fully featured virtual instruments and sampler hosts available. For more on virtual instruments and samplers, refer to Chapter 17 for a full description. Logic comes in two versions: Logic Express and Logic Pro. Logic Express is the basic version and Logic Pro is the flagship application. What you gain with the Pro version are more plug-in effects, more tracks, and generally more features. **FIGURE 6-4** is a shot of Logic in action.

Cakewalk: Sonar

Sonar is an immensely popular workstation made only for PCs. Cakewalk has been making music programs for years and one thread that runs through each of its products is ease of use. On the PC side, it's one of the most popular home studio products out there. A clean, uncluttered interface makes the connection between user and musician seamless and fun. Excelling at both MIDI and audio, Sonar is one of the best PC applications available, and it has a wide, loyal user base. You also get automation, a generous selection of plug-in effects, and a wide variety of MIDI editing techniques.

QUESTION?

On a computer, how many audio and MIDI tracks will I have?
Since these programs rely on your computer, it's hard to answer that! For MIDI, you'll get more than you can ever use. With a fast, modern computer you can get more than thirty-two audio tracks. Using effects will reduce the number of tracks you can use. This also can vary depending on the particular program you are using.

Which One Is for You?

The popular software products discussed in this section do the same basic job—they all allow you to manipulate MIDI alongside audio in a non-linear manner. They differ only in presentation and organization. You owe it to yourself to look into each product. You'll find many demos are available online as well as at your local music store. Get some hands-on experience with the software before you purchase it. Ask around; ask friends. Online, there are dedicated support forums for each product. Read the entries and look into all the details. Buying a program is a major commitment, so do your homework.

If you plan to collaborate with other musicians you know, or other people in your band, it would pay for you all to have the same system. (And if you buy the same system as someone you know owns, you might just be able to get a little tutorial from that person as well!) If you have aspirations of becoming a professional studio owner or attending engineering school, see what the current standard is—for most, it's still Pro Tools. Whatever you decide to get, you'll learn and do well with it!

Chapter 7

MIDI

MIDI has been around for more than twenty years now, yet few musicians know what it can do for them. With the swift technological changes that have occurred in recording technology, MIDI has remained virtually unchanged since it was first implemented as a standard way for computers and keyboards to communicate.

The Basics of MIDI

MIDI stands for "musical instrument digital interface." MIDI was conceived in the 1980s as a way for computers to control keyboards, and for keyboards to control each other. MIDI is the modern equivalent of a player piano. Player pianos used punched-out paper that the piano read and self-played from. MIDI is an electronic language that synthesizers turn into music. Even though keyboards are capable of producing some very complex sounds, the manner in which they produce sound is very simple.

Here is what a MIDI keyboard sees when you play any single note:

- **Note On:** Turn on the note you've pressed (for example: Note On, C4)
- **Volume:** Set the loudness (0–127)
- **Note Off:** Stop the sound

MIDI has many other commands for controlling other parameters, which we don't need to get into here. The essential information is small and easy to transmit from keyboard to keyboard, or computer to keyboard. As you can see, it's pretty simple to have a keyboard controlled with MIDI. It's very much like the old player piano that reads rolls of punched-out paper.

Even better is that the commands sent to control MIDI are very small, just small lines of textlike code. Even twenty years ago computers could handle complex MIDI. MIDI is not just for keyboards anymore—samplers, virtual instrument plug-ins, sound modules, guitar synthesizers, and drum machines can all utilize MIDI technology.

In the early days of MIDI, keyboard players and pianists were the largest group of MIDI users. One of the joys of MIDI was that it could record full performances exactly as you played them, directly into the computer. It allowed a piano player easy access to computer score writing and sequencing. Since only commands were recorded, not audio, performers could easily control the performance on even the most minute levels! They could alter the timing of individual notes by milliseconds. You could, and many did, go nuts making everything "perfect."

MIDI also worked in reverse—musicians could compose something on a computer either by entering notes in a score program like Finale or Sibelius. MIDI was used to bring the music to aural reality. The music programs sent out

MIDI commands for the keyboards and sound modules to play back the on-screen score. MIDI allowed composers to hear pieces in a semirealistic fashion, especially if groups weren't present to perform them.

Drum machines are also MIDI capable, and many drum loops and beats can be composed on the computer and sent to the drum machine for playback. **FIGURE 7-1** shows a MIDI keyboard.

FIGURE 7-1

▲ MIDI keyboard

How MIDI Is Used Today

In some ways, MIDI hasn't changed at all. There are still legions of keyboard players who use computers and their sequencer software programs to build multitrack arrangements. Composers still use MIDI to play back performances.

In the last few years, samplers and virtual instruments have changed what is possible to produce in a home studio. The sound quality of the modern samplers and virtual instruments are astounding to say the least. With early MIDI synthesizers, it was easy to tell that the instruments were not real. Nowadays, with the quality of samplers and virtual instruments, it's getting hard to tell what's real and what's not!

Samplers and virtual instruments are all MIDI-controlled. So to really take advantage of them, you have to have some way of interfacing with MIDI. This doesn't mean that keyboard players are the only ones who get to enjoy this!

FACT

Many of the virtual instruments and samplers require only one or two notes to activate great loops and sounds; so being able to play a keyboard well is not required. And since MIDI can be recorded one note at a time and edited together later, anyone can get in on the fun. You don't need to play fast, or even play in time! Even the one-finger piano player can get into the MIDI world.

Guitar players can use guitar synthesizers to access the exciting world of MIDI. Brian Moore Guitars manufactures the iGuitar, which is a traditional guitar with a built-in thirteen-pin MIDI pickup inside the guitar. Simply attach the guitar to a converter box and your guitar playing is instantly converted to MIDI. Guitarists can finally easily sequence, play soft synths, and produce notation.

For the more adventurous guitarist, there is an instrument called the Ztar, a keyboard with keys in the shape of a guitar neck, which is played like a guitar but it has no strings, only buttons.

For those who play a little keyboard . . . you might benefit from a "little" keyboard. Many companies manufacture small desktop keyboard controllers (shown in **FIGURE 7-2**). These controllers contain no sounds; they transmit only MIDI messages to the computer. The products are very affordable (many around $125) and only contain two to three octaves of keys. It's a great way to utilize MIDI in your setup. You can, of course, purchase larger keyboard controllers for extended range.

FIGURE 7-2

 Small MIDI keyboard controller

Essentials for Using MIDI

To get started with MIDI, you need three things. First, you need a device capable of sending out the MIDI messages, such as a guitar synthesizer, a keyboard, an electronic wind instrument/controller, and so on. Second, you need a hardware interface to get MIDI into the computer. You will probably want a computer so you can take advantage of MIDI's powerful editing ability. The higher-end "keyboard workstations" like the Kurzweil and Korg products have internal sequencers built into their keyboards; however, most users take advantage of a computer's flexibility and power. If you go with a computer, the third thing you will need is a computer program to perform the sequencing. All the major digital audio workstation programs (DAW) have integrated MIDI sequencing. (See Chapter 6.)

Interfaces

Computer MIDI interfaces are inexpensive and easy to use. They are available as part of PCI cards, USB, and Firewire interfaces. Many of the newer computer recording interfaces now include MIDI ports. The minimum you will need is one input and one output. You will need one input for each piece of gear you want to use. But don't worry—if you have a lot of MIDI gear, you can buy interfaces that handle eight or more inputs!

What a MIDI Cable Can Carry

One simple MIDI cable can transmit a lot of data! Most of the modern keyboards and sound modules are able to receive sixteen channels at once. That means you can have your sequencer send up to sixteen different tracks of MIDI to your keyboard and your keyboard will be able to play back all sixteen different tracks from just one MIDI cable.

Not every keyboard is capable of receiving and playing back multiple channels at one time, so if that's what you want, you need to look for a keyboard that is "multitimbrale." A device is multitimbrale if it can play back more than one "timbre," which is French for sound. Most modern products are multitimbrale.

Recording MIDI

While it is possible to record sixteen sounds at once, most likely you'll record one at a time as you start out. The first thing you need to do is tell the program what channel to listen for, so make sure the keyboard is transmitting on the correct channel. Since every keyboard implements its menus and commands differently, you should check the documentation that came with your keyboard for this step. For those using a controller keyboard, most will transmit on a fixed MIDI channel, many solely on channel one. Once you've set up both the computer and the keyboard to talk together, you'll be able to record-enable the track and see that MIDI is being received. **FIGURE 7-3** shows a screenshot of a record-enabled MIDI track in Cubase software set to receive on channel one.

FIGURE 7-3

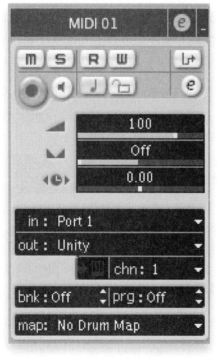

◀ MIDI channel
Screenshot used by permission of Steinberg Media Technologies.

Once you've recorded the MIDI information, you can edit to your heart's content. Repeat the steps to build up as many separate tracks as you need. If you are recording more than one channel at the same time, set up multiple tracks and assign each one to a different MIDI channel, and record-enable all of them.

Playing Back MIDI

Just like recording MIDI parts, playing back MIDI requires that you assign each musical part to a different MIDI channel, one sound per channel.

FACT

"One sound per channel" doesn't mean monophonic, or one note at a time. One channel can be a piano patch playing many chords. Each keyboard of sound module will have a set "polyphony" number that sets how many notes can be played back at one time regardless of what channel it's on. Modern keyboards and sound modules allow for high polyphony, usually around twenty-four to forty-eight notes at once.

When playing back the recorded MIDI tracks, you can assign the MIDI tracks to specific sounds within the keyboard or sound module. For instance, you can have a string orchestra play your piano chords. Each separate track allows you to change the patch number and channel it transmits its MIDI data on. So, for example, you can record a simple part with your keyboard controller and assign it to a piano sound on your sound module or sampler. In the software program, you set the transmit channel to match the channel that your desired sound is located on the keyboard or sound module. Some keyboards and sound modules require you to set up the sounds on the unit itself, channel by channel. Most modern units allow the MIDI software to perform a "control change" and switch presets for you. Check your user manual to see if your equipment has this capability.

You should now be able to get started with MIDI in your studio. Remember, the beauty of MIDI is the control it gives you over your performance. Learn to embrace the advanced editing tools that your software of choice affords you.

Chapter 8

Setting It All Up

All that gear you have is no good to you unless you know how to set it up. It's safe to say that every stage in the recording process is important. Setting up your gear is no different. This is an essential step where many people make mistakes. You might even learn some tricks to simplify your life. A good setup goes a long way.

Cable Types Explained

There are a few types of cables that you'll have to deal with in your studio. So you know what they do, here is a breakdown.

Quarter-Inch Cables

The most common cable is the quarter-inch cable. It's the familiar "guitar cable" that we all know and love. These cables are used for plugging guitars and keyboards to amplifiers, amplifiers to mixers, and mixers to speakers, just to name a few. However, all quarter-inch cables are not created equal. There are specific types for specific uses. The most common is the instrument cable. The instrument cable is used for plugging instruments (guitars, keyboards, samplers) into amplifiers and other sources. The cable carries the mono (one-sided) signal to and from any source you choose.

Instrument cables are also commonly referred to as tip sleeve cables. If you look at the connection end, you'll notice that the tip is separated from the rest of the connector with one plastic spacer. The spacer separates the hot side of the cable that carries the signal from the side that carries the ground.

The other type of quarter-inch cable is the speaker cable. Speaker cables are very different from instrument cables. A guitar or a keyboard puts out a very small amount of power, so you might get some noise, which no one likes to hear in recordings. Instrument cables contain a shield inside the cable to help keep the noise down. But speaker cables push more power through their cables than instruments do. Because of this, they don't need a shield, due to something called signal-to-noise ratio.

Signal-to-noise ratio is *critical* to understand. The higher the level of a signal, the less you hear the noise that is present. This is why you should never record low levels. Noise, while it's annoying, can usually be covered up by a full, loud signal.

Speaker cables shouldn't be used for instruments and vice versa. The packages clearly state what the cables are used for. When in doubt, ask for help.

FIGURE 8-1

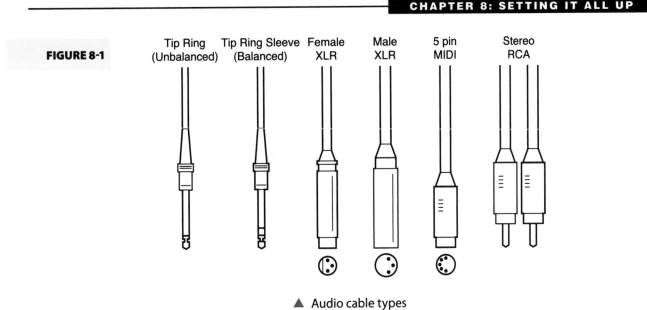

▲ Audio cable types

FIGURE 8-1 shows all the connector types: quarter-inch TR, quarter-inch TRS, XLR, MIDI, and RCA.

Tip Ring Sleeve Cables

The tip ring sleeve (TRS) is a quarter-inch stereo cable. Pull out any pair of stereo headphones that you own and look at the plug. Notice how it was two plastic spacers? The quarter-inch tip sleeve (instrument) cable has only one. The extra spacer on the tip ring sleeve is there to accommodate another signal in the cable. Stereo cables carry two separate signals: left and right.

In a studio, a stereo cord is used for two purposes. First, it connects an effect into a mixer or recording device. This type of cable is called an "insert." Insert cables have the stereo connector on one side and two mono cables on the other side. The stereo line splits the signal so that you can have an input and an output to an effects processor. (If that sounds confusing, don't worry, it's covered in more detail in Chapter 15.) In short, if you're going to use external effects, you'll need to own a few stereo insert cables.

The second use for a stereo quarter-inch cable is for a balanced signal. Balanced cables have the stereo plug on both sides. What does balanced mean? Basically, the cable copies one signal to the two internal wires and performs extra shielding and other electrical magic to cut down the noise.

Balanced cables are used for microphones and whenever cords lengths are very long. This helps cut down on the noise that longer cables usually have.

FACT

Balanced cables can be used only if your mixer or recording device supports them. It will be clearly stated in your manual if you can use a balanced signal. You find balanced connections on better recording equipment. Microphone inputs are *always* balanced.

XLR Cables

XLR is the standard microphone cable, and the letters stand for the three signals carried in the cable. X is for external, which is called the ground, L is for line, and R is for return. The cable is round and has three prongs on one side (the male side) and three sockets on the other (the female side). Microphone cables are always balanced cables. XLR cables are sometimes used for connecting mixers to recording devices, but their most common use is for microphones.

RCA Cables

RCA cables, invented by the RCA Corporation, are another type of unbalanced cable. They typically are two mono cables that run together and split off into two ends. RCA cables are commonly used in home stereo equipment. In the recording studio, RCA cables are used to hook up a tape machine to a mixer and recording interfaces to some computers.

Digital Cables

Digital audio signals fall into three categories: The S/PDIF digital cable carries one stereo signal digitally on an RCA cable; the AES/EBU cable carries the same stereo digital signal, this time on XLR cables; fiber optic cables are used for audio. Digital cables can transmit a stereo pair, or in the case of ADAT light pipe, eight signals at once! Digital connectors are very common on recording equipment today—even lower cost ones.

MIDI Cables

MIDI cables are simple, and there is only one type: five-pin MIDI. You can't possibly buy the wrong one. You'll need one cable for each input, and one cable for each output.

Patching, Levels, and Monitoring

There are a few other vital areas to talk about that don't really fit anywhere else here. These include essential gear like patch bays, important information about setting proper levels, and setting up your studio monitors and monitoring setup. Understanding these will really help complete your knowledge of home studios and what components go into them.

Patch Bays to the Rescue

If you have a lot of gear that gets plugged in and replugged often, you're going to love a patch bay. A patch bay (shown in **FIGURE 8-2**) doesn't look like much. But don't let looks fool you; this is one powerful piece of gear.

FIGURE 8-2

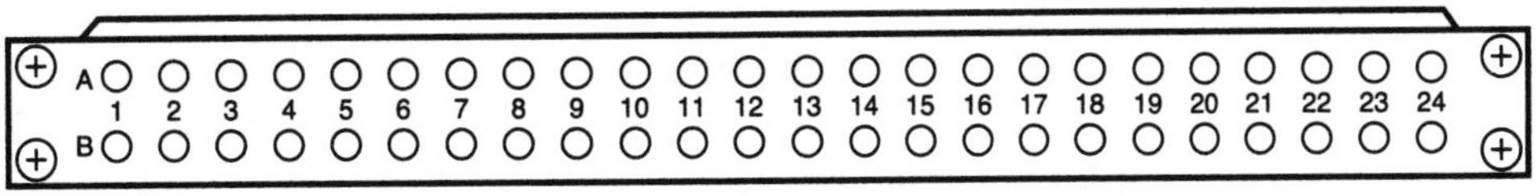

▲ Audio patch bay

On the front panel are lots of input jacks; on the back panel are lots of matching jacks. Instead of crawling around the floor, repatching (reconnecting) all of your gear, you can leave it permanently plugged in to the back of the patch bay and use the front to make connections. Patch bays are also great for mixers that have rear connections because you're saved from the headache of crawling around or lifting up the mixer every time you need to patch something. If you have a small studio, you might never need one, but as you grow in size a patch bay is essential. You won't find a professional studio without several patch bays.

Understanding Input Levels

Setting a proper recording input level is essential. A full signal is best. You always want to record as loudly as possible without clipping (overloading the input, causing distortion) the input or preamp. Every recording device, even the cheap tape 4-tracks, has a meter that shows you the level of an incoming signal. **FIGURE 8-3** shows what a typical meter looks like.

FIGURE 8-3

Left		Right
–O–	28	–O–
–O–	10	–O–
–O–	7	–O–
–O–	4	–O–
–O–	2	–O–
–O–	0	–O–
–O–	2	–O–
–O–	4	–O–
–O–	7	–O–
–O–	10	–O–
–O–	20	–O–
–O–	30	–O–

◄ Mixer levels

Usually the displays on a meter light up green at the bottom and red at the top. Just like a stop light, red on the meter means stop. Why? Because you're clipping, and you need to adjust the level. Your goal is to stay green most of the time and let your loudest parts stay away from the red, although it's okay to get close to it. By keeping the levels high when you record, you'll maximize the signal-to-noise ratio and drown out any noise you might have. If you record too softly, you'll have to boost the signal during mixing, and that boosts the noise along with it. No matter how loudly you record a signal, you can always turn it down later when you mix.

Setting Up Your Monitors

If you choose to monitor through headphones, setting up your monitors is as easy as plugging them in and placing them over your ears. If you're going to use monitor speakers, setting them up is almost as easy. First, place the monitors level with your ears; placing them on the floor just won't do. Face them slightly in toward each other so that both speakers focus their sound at the middle of your head. If you were to draw a line from your head to one speaker, to the other speaker, and back to your head, the line should form an equilateral triangle.

FIGURE 8-4

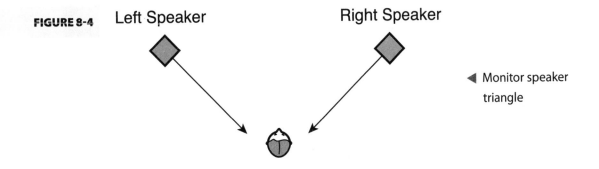

Left Speaker   Right Speaker

◀ Monitor speaker triangle

Monitors are also referred to as "near field monitors" because you shouldn't be too far away from them. You should sit less than 5 feet away from your monitors (3 or 4 feet away is optimal), and set up the monitors that same distance from each other as well. Use balanced cables, not instrument cables, to hook up your mixer or recording machine to your speakers.

Understanding the Signal Chain

The signal chain is the path your sound takes when it is recorded. Plugging your microphone into a mixer and plugging the mixer into a recording device is a common signal chain. Signal chains can be very simple or quite complicated depending on your setup. Here are two key elements to understanding signal chains:

- **Gain:** This is the recording engineer's term for the volume or loudness of a signal. Every time you turn up a guitar amplifier, move a mixer's slider up, or turn up your headphones, you're adjusting gain.
- **Gain stage:** This is any device that changes the volume of a device. Mixers, microphone preamplifiers, and input level controls are all gain stages.

Gain and gain stages are really important to recording great signals. Proper gain is important to getting a clean signal. Every time your signal passes through a gain stage, it is affected. The fewer gain stages you pass through, the better. It's possible to clip a sound in one gain stage and turn it down in the next, resulting in a terrible signal—a low volume signal that clips. Plus, excessive gain stages can introduce noise in your signal, so less is better. You use gain stages to set the level of the instrument to a good, full level without distortion. Distortion is caused by adding too much gain and overloading.

There is a credo shared by many in the recording business: garbage in, garbage out. If any part of your signal chain is weak, that weakness will come out in the recording. Bad microphones will still sound bad when played back on expensive recording systems. Noisy crackling cables will crackle on the recording. Clip an input and it will get recorded clipped. Recording isn't a magic wand that magically cures problems. There certainly are some tricks of the trade that we can share, but you get what you put into it. Recording is very honest: Whatever you give it, it spits back at you.

ALERT!

If you're just starting out or have never recorded before, your first recordings might depress you a bit; you might not sound as good as you think! Little problems in pitch and rhythm will be very clear when you listen to your recorded music. Recording is, in a word, honest. Many people find it a great tool for learning their strengths and weaknesses.

Isolating Sources of Sound

Sound travels in waves, and it's always loudest directly in front of the sound source—the amplifier, the guitar's sound hole, the mouth of a singer, and so on. Once sound is released from its source, it can reflect and bounce all over a room. It's very hard to control where the sound goes.

Being able to separate sound is very important in recording. That's why isolation booths are a staple of recording studios. Being able to isolate sound is even more critical to the home studio owner because you don't have the luxury of using soundproofed rooms and isolated sources. So what can you do?

Banish It

Let's set up a scenario: You are recording a guitarist, and the amplifier is right next to the recording device. Unfortunately, because your room is small, you don't have much of a choice about placement. The guitar player, like most guitar players, loves to play his amplifier loud. So, you hook up a microphone and turn on your headphones or monitor speakers. Yet you can't hear anything but the roaring guitar amplifier 3 feet from you. How are you going to listen for microphone placement, clipping, or anything else for that matter? Well, in short, you won't. You need to get the offending sound (or player, whichever comes first) isolated so you can do your work.

The easiest way to isolate is to put the amplifier in another room and close the door. The guitarist can monitor his sound through headphones along with you. This way you'll be able to hear what the recorder is hearing and not what the amplifier is forcing you to hear. Closets work well for this because they are small, and if they contain jackets and sweaters, the clothes will also absorb some of the sound. The bathroom can isolate sources pretty well, too! And it's also a great place to record sources, especially acoustic guitars, acoustic instruments, and vocals. Since most bathrooms are tiled, the sound is very reflective and live, which can be really good for adding some ambience and reverb to your recordings. You'd be amazed at how many home studios rely on the acoustics of a bathroom for recording sources.

FACT

A great baffle you can use at home is a mattress. Positioned a few feet in front of an amplifier or a drum set, the mattress will absorb and block some of the sounds from overpowering you. It's not a perfect solution since sound reflects to other places, but it can *really* help you out if you're in a pinch.

Using Baffles

A baffle is an object that blocks sound. It's also referred to as a "gobo." If you don't have the option of isolating sound to another room, a baffle might do the trick.

FIGURE 8-5

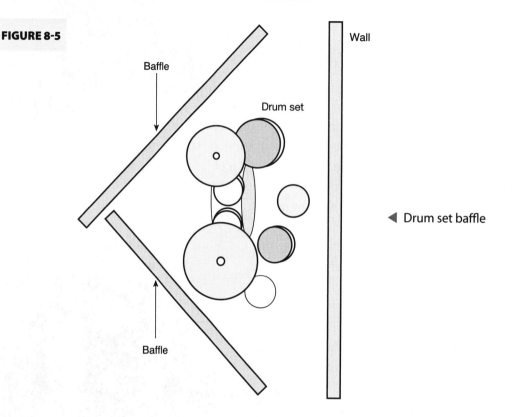

Wall

Baffle

Drum set

◀ Drum set baffle

Baffle

Baffles are often used onstage around drum sets (see **FIGURE 8-5**), typically made of Plexiglas. These help to keep the drum sound from traveling out on the stage and getting picked up by other microphones. If you plan on recording a live band performance and you wish to retain any hope of mixing signals separately, you'll need baffles because sound has a nasty habit of flying everywhere if you don't control it. The drums will bleed into the guitar microphone and so on, making it hard to separate sounds and mix independently. Putting baffles between instruments can do the trick.

Soundproofing

You can do two main things to defeat sound in its tracks:

- **Create dead air:** Dead air traps sound and doesn't let it escape. You can trap air with a mattress, cushy furniture, or suspended (drop) ceilings.

- **Add mass and density:** A massive or dense object stops sound in its track. For example, you can scream as loud as you want at a concrete wall, but you won't be heard on the other side. Because sound is produced by vibrating objects that push the air around them as sound waves, a substance with little flex (like concrete) won't vibrate well. Less vibration equals less sound.

The best way to soundproof a home studio is to build a room within a room. After the normal drywall is hung, add sound batting, which is an absorbing material. Then hang another layer of drywall a few inches out from the original wall or ceiling. This creates a space between the walls to trap the sound, and it also adds to the mass and density of the wall. This is a common practice in professional studios, too. There are many books and resources available on soundproofing, so check the Web, your local music store, and the public library near you.

You can also isolate the amplifiers by building an isolation box for them. By sealing each amplifier in a box, you create a mini–isolation room and prevent sound from escaping. You'll also need to insulate the inside of the box so the sound gets trapped there. Make sure the box is big enough to accommodate the microphone, and cut a small hole in the box to get the microphone cable out. Once you close up the box, you should be able to crank the amplifier and have little sound leakage. You'll never eliminate *all* extra sounds this way, but you can reduce the sound by almost 90 percent.

Acoustics

The science of acoustics is far too vast and complicated to get into any depth here, but some basic understanding will help you in setting up an optimal space for your studio, wherever it might be.

Sound Absorption and Reflection

As sound travels through the air and interacts with surfaces, it's either absorbed into the material or reflected. Absorption is typically the least of your worries; most of the time it's a blessing in a home studio. Reflection, on the other hand, can be a problem. In a big space, reflections can have an ambient reverb, which is a good thing. Reflections are an essential part of a live concert hall, especially a classical hall, because they produce a natural reverb that is pleasing; reverb processors try to imitate the sound of a big room's natural reflections. But in a home studio that's set up in a small space, reflections can be as loud as the original signal, making it hard to mix and hear the real signal. If you have a space that is overly reflective, the easiest thing you can do is add some materials to absorb the sound and stop the reflections.

Foam for Your Walls

Foam is one of the greatest sound absorbers. Acoustical foam, a special foam designed for studios and sound professionals, is a textured material with raised surfaces and depressions. It does a really good job of cutting down the reflectivity of a room, and is great when mixing with speakers. When you record, the sound of the room is recorded as well; this is called ambience. Certain microphones pick up less ambience than others, but there's always *something* that gets picked up. If your room is overly reflective, you might get too much ambience in your signal. Acoustical foam can help tame those sources.

You can purchase foam from most major music stores or through online catalogs; it is sold by the sheet and can easily be mounted nondestructively to the wall. Acoustical foam does a bad job of soundproofing; so don't try to use it in this capacity.

The Direct Approach

If noise is a problem where you live, try as hard as you can to record direct sources. Electric guitar doesn't have to be miked; there are plenty of processors that do a great job of emulating that sound while the guitar is directly plugged in. The Line6 POD is a famous example of such a processor, and so is the Behringer V-AMP. With certain instruments, such as the drums, you have no choice but to use a microphone. The more direct sources you can use, the simpler your noise problem will be. It's possible to record all direct instruments (guitar, keyboard, bass, drum machine) and mix on headphones, thus creating zero audible sound.

Common Mistakes

It's a pretty safe bet that your first recordings will have some problems, so we should go over some very common ones and show you easy ways to fix them.

Recap of Signal Chains

Keep your signal chains simple. Don't use excessive wiring if it's not necessary—the simpler the chain the better. This is especially important when dealing with gain stages; you have to be really careful about levels. Any device that can raise and lower your gain, such as a mixer, preamplifier, or input trim control, must be used carefully. It's easy to make mistakes like using two gain stages when only one will do, possibly adding noise or easily clipping the signal. Remember, once you clip a signal by overboosting it, your take is destroyed. You can't mix out a clipped track. So be careful!

Ground Loops and Buzz

So you have some noise? Where is it coming from? There are a few very common buzzes, and most are easily remedied. Most buzz comes from bad grounding, which can create what is known as a ground loop. Most modern houses utilize three-prong electrical outlets; the third plug is a ground, which draws excess electrical charges away from a power source. If an electrical outlet is not grounded properly, every device plugged into it might buzz. This can include your amplifiers or even your recording devices. Buzz is annoying and can ruin a recording. Fortunately, there are a few things you can do.

First, try the easiest fix: Plug the source into different power outlets. Amazingly, some outlets buzz less than others. If that doesn't work, purchase a power conditioner, a device that regulates the type of power that comes out of an outlet, making sure the power is regular and clean. If that doesn't work and you have a really bad buzz problem, call an electrician to look at the wiring in your house. This can be costly, but might be a viable option if the noise is rendering your expensive gear useless. Direct boxes are also susceptible to ground loop buzz. Thankfully, most direct boxes come with a ground lift switch that eliminates the buzz when activated. If you're looking to purchase a direct box, look for one that contains a ground lift.

Noises of All Kinds

Ground loop is not the only kind of noise. Electrical equipment, especially delicate equipment such as microphones and recording devices, is prone to interference. Interference from radio signals, televisions, and magnetic sources can provide problems. It's not uncommon for a PA system to occasionally pick up a radio station or cordless phone signals, and you might run into these problems in your home studio as well. The most common problem that home studio users face is from magnetic interference. If you've ever put a magnet near a TV, you've seen the color go haywire. Studio monitor speakers contain large magnets as part of the speaker. If you run a computer-based studio with a tube monitor, the speakers could interfere with the signal in the computer (not to mention that magnets should be kept away from hard drives). Look for monitors that are magnetically shielded. Many monitors are designed this way due to the popularity of computer recording, but you should double-check to be safe.

Chapter 9

Microphones

There are so many brands, sizes, and shapes of microphones to choose from. This makes it hard for a novice to walk into a store and buy the right one. While there might be hundreds of microphones, there are actually just a few types to choose from. Once you get a handle on them, you can make informed decisions.

Getting Started

To record any acoustic instruments, such as acoustic guitar, piano, wind instruments, drums, or voice, you'll need some microphones to pick up sounds and convert them to electric signals. While all microphones do basically the same thing, not every microphone is right for every purpose. Just pick up a music catalog and look through the microphone section, and you'll see many shapes, sizes, brands.

If you're the only one recording (you are working alone and playing all the instruments), you should have one microphone for every instrument you want to record. You can always buy more at a later time if you need to. If you plan on having groups record live or making some money from your studio, you'll want a wide selection of microphones to cover any situation that comes up. But how do you know what you need?

Polar Patterns

One essential aspect is what direction a microphone hears. To understand this, you must understand a little about polar patterns, because they are essential to picking the right microphone. Polar patterns are easiest described when looking at one. **FIGURE 9-1** shows the polar patterns you will encounter when microphone shopping.

Around the outside of the circle are degree markings. These degrees signify the origination of the sound in relation to where to the microphone is pointed. Simply, the microphone is pointed at 0 degrees. The shaded areas in the circle show direction and scope of what the microphone can hear and what it can't. In the case of the cardioid microphone, you can see that it hears best what's directly in front of it, at 0 degrees. If you were to stand behind a cardioid pattern microphone, at 180 degrees, the microphone would be basically deaf to anything you say. By looking at polar patterns, you can tell the direction and scope of what the microphone will hear and what it won't. Out of the five polar patterns you see in this figure, there are only three main polar patterns (cardioid, omnidirectional, and figure-eight), and the remaining two (hypercardioid and supercardioid) are subpatterns.

FIGURE 9-1 ▼ Microphone polar patterns

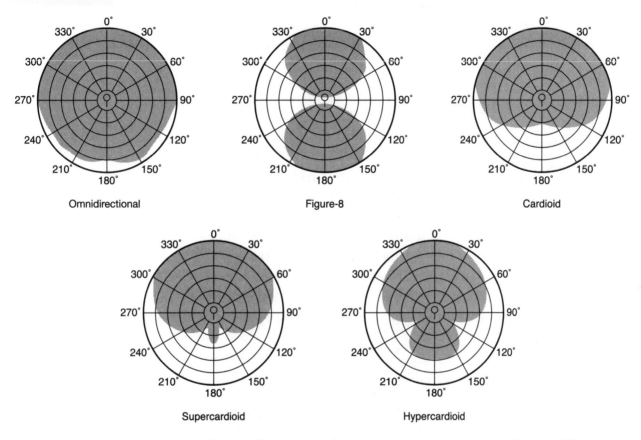

As we said, a cardioid microphone hears sounds in front of it best. The subtypes, hypercardioid and supercardioid, simply change how much it can hear behind. If you look at **FIGURE 9-1**, you will see that both hypercardioid and supercardioid have very little shaded areas behind them. You can minimize microphone bleed by close-miking an instrument with a supercardioid microphone, since it rejects sounds from behind it so well. A supercardioid will only hear what's right in front of it. If you use this microphone pointed right at your sound source, you won't experience much bleed from other instruments in other directions.

An omnidirectional microphone hears everything from all sides. You can see in its polar pattern that the shading is all around the microphone. Omnidirectional microphones are great room microphones and work well with large groups.

A figure-eight microphone hears sounds only directly in front and directly in back of it, rejecting much of the sounds from the sides. Figure-eight microphones are great for recording two vocalists on one microphone, or for recording between two instruments, blocking out the sides.

Frequency Response

Frequency response is another factor to consider in selecting a microphone. Each microphone hears sound differently. Microphones will boost or cut certain frequencies across their range, which is what frequency response means. Many microphones aim for a flat response, which means there is little boosting or cutting of frequencies. With a flat response, you get as accurate a representation of the source sound as possible. There are boundaries to how low and how high microphones can hear. The lowest frequency that most microphones can hear is 20Hz, and the highest frequency is 20kHz. You will see in product literature and advertisements these numbers (20Hz to 20kHz response). **FIGURE 9-2** shows a frequency response chart for a dynamic microphone.

FIGURE 9-2

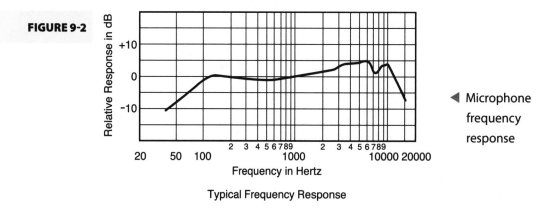

Typical Frequency Response

◀ Microphone frequency response

However, not all microphones go down to 20Hz, and not all go up to 20kHz. So, why is this information important? Consider the boundaries of sound. The lowest note on a piano registers at 27.50Hz, which is about the lowest sound you can really imagine recording. The highest sound on a piano clocks in at around 4kHz. There are, of course, higher frequency sounds. We won't get into this here, but all sounds spread into many frequencies, and there are always higher overtones present. Microphones need to hear pretty high for our ears to accept the sound as "natural."

FACT

The loudness of any sound is measured by the level of sound pressure present, measured in decibels (dB). An airport runway is about 120dB and a normal conversation is about 60dB. Loud amplifiers, screaming singers, and certain drums can put out extremely high sound-pressure levels. If the microphone you choose can't handle the sound-pressure level, you can damage or distort the microphone over time, rendering the signal unusable.

Dynamic Microphones

Of the three main types of microphones available, let's start with the microphone most people are familiar with, the dynamic microphone. Dynamic microphones are the microphones you've most likely come into contact with if you've ever played on a stage, sung, or spoken into a microphone. Dynamic microphones are the "ice-cream cone"–shaped microphones that we all know and love. What makes this microphone "dynamic" is the way in which it picks up the sound and translates it into electricity. It has a small diaphragm made of plastic, usually Mylar (a type of plastic), that is placed in front of a coil of wire, called the voice coil. The voice coil is suspended between two magnets. If you remember from high school physics, when you move wire between magnetic fields, you can induce current. As sound hits the diaphragm, the voice coil moves and induces current. The current is fed down the microphone cable into your mixer or recording device. Voilà, sound!

Dynamic microphones are great for vocals, miking amplifiers, and close-miking drums. Dynamic microphones are durable, well constructed, usually last a long time, and—the best part—cheap. You can get a great microphone for under $100.

Dynamic microphones hear sound in cardioid patterns, which means they hear sound only directly in front of them. This makes them great for close-up miking situations, such as vocals, amplifiers, and drums, where you want nothing but the sound coming from the source with little or no ambience. One of the most important points about the dynamic microphone is its ability to handle extremely high sound-pressure level (SPL). As a result, you can use them in loud situations. However, dynamic microphones don't have full frequency response, which is helpful in many situations, and all are a bit different in the frequency levels they respond to. For example, the AKG D112 microphone is made for miking bass drums and bass guitars. Its frequency response emphasizes the bass frequencies and falls off at the high range, where there is little signal from a bass drum.

FIGURE 9-3

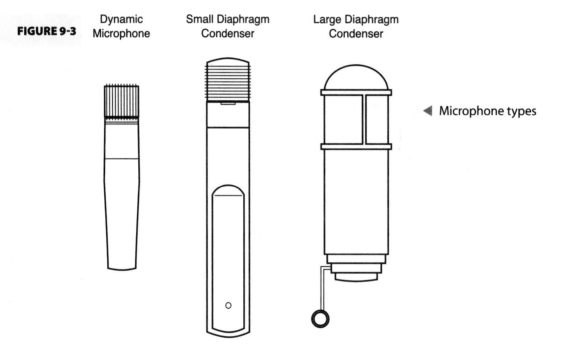

Dynamic Microphone · Small Diaphragm Condenser · Large Diaphragm Condenser

◄ Microphone types

Condenser Microphones

Condenser microphones are commonly found in recording studios but not often on stages because of the different way in which they pick up sound. Condenser microphones use a diaphragm, just as the dynamic microphone does, but instead of sitting in front of a wrap of coil (the voice coil), the diaphragm sits in front of a stationary plate of metal called the back plate.

The diaphragm in a condenser microphone is always made of thin metal or metal-coated plastic. Both the diaphragm and the back plate have polarized voltage applied to them. As the diaphragm moves back and forth in relation to the stationary back plate, a very small current is produced, which is your signal.

Condenser microphones are often classified by the diameter of their diaphragms. Obviously, large-diaphragm microphones are larger in size than small-diaphragm microphones. Large diaphragms are more sensitive to low level sounds, than small diaphragms. Small diaphragms can handle louder overall sounds, however. Condenser microphones can range from $100 to many thousands of dollars. The good news is that you don't have to spend a lot to get a good-sounding one.

FACT

Another type of microphone is the ribbon condenser. Instead of using a traditional diaphragm as the dynamic and condenser microphones do, a ribbon condenser uses a very thin and flexible ribbon to pick up the sound. Ribbons are very fragile and can be damaged with mishandling. Because of their high price and fragility, they are much less common in home studios. Recently, budget-priced ribbon condensers started to appear on the market. Even the budget ribbon microphones are much more expensive than other budget microphones, but they have a unique character to their sound that is worth investigating.

Condenser microphones are used for everything that dynamic microphones can't do. Condensers are more delicate and more expensive than their counterpart dynamic microphones. What condensers have going for them is that they re-create the sounds they are given very accurately. Condensers usually have flat and wide frequency response, hear all frequencies, and tend not to boost or lower any particular frequency. This is why condensers are the microphones of choice for acoustic instruments such as piano, winds, strings (including guitar and bass), drum overheads, and vocals.

As for polar patterns, some condensers can utilize all polar patterns. This makes them ideal when you have to pick up a whole room with one microphone using an omnidirectional polar pattern. Couple this with the fact that some condenser microphones can switch polar patterns, and you've got one flexible microphone. Most condenser microphones can't handle super-high sound-pressure level, so don't replace all your dynamic microphones with condensers. On certain condenser microphones, you might get a pad switch that lowers the microphone's output by 10dB or more in order to better handle louder sounds.

Condenser microphones do a much better job than others at capturing a true spectrum of sound, from its lowest to its highest pitch. However, condenser microphones are more sensitive and delicate, and typically can't handle high sound pressure. You can actually fry a good condenser with high SPL if you're not careful!

ALERT!

Repeated exposure to high SPL can cause permanent hearing damage. Don't crank the sound in your studio and be especially careful with headphone volume. You get only one set of ears.

Microphone Extras

You mean there's more? Yes, you're not out of the woods just yet! The good news is that if you've hung on through all this, you really know something about microphones and what makes them tick. This will be invaluable to you as a home studio owner. So here's the rest.

Stands

A microphone doesn't just lie on the floor—you need to use a microphone stand to secure and position it. This is a rundown of the basic kinds of stands you'll encounter:

- **Standard:** This is your garden-variety microphone stand. It has a wide base and allows for up-and-down positioning of the microphone only.
- **Low Profile:** This is just like the standard microphone, except it's very short. It's great for bass drums and amplifiers that sit on the floor. It provides up-and-down height adjustment only.

- **Low-Profile Boom:** This is a short stand coupled with a boom arm for much greater maneuverability on all directions, not just up and down.
- **Standard Boom:** This is a tall stand with a boom arm for maneuverability in many directions. It's great for vocals, drum overheads, and much more. This is a good all-purpose stand.

You can't go wrong with boom stands because they can work in practically any situation. If you microphone a lot of bass drums or amplifiers that sit on the floor, you'll need the low-profile stands to get down low, although you could also lean a boom stand down.

Shock Mounts

Many condenser microphones come with a strange-looking apparatus called a shock mount (**FIGURE 9-4**), which is a web of elastic string that encloses the microphone and is secured to the microphone stand.

FIGURE 9-4

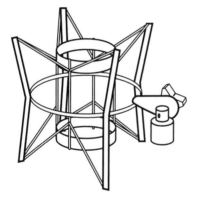

◀ Microphone shock mount

The shock mount tries to isolate and cushion the microphone to prevent it from picking up sound traveling up the microphone stand. Remember, sound can be transmitted through objects, and microphone stands are no exception. Condenser microphones are very sensitive and if you accidentally stomp a foot on the floor, or worse, hit the microphone stand, the microphone will pick up the thump and ruin your recording. If you place your microphone stand on a bare floor, the microphone stand might vibrate due to other frequencies in the room, especially other low-end instruments. You don't have to use a shock mount; but if your microphone comes with one, you should use it.

Pop Filters

If you plan to record vocals or spoken word, you need a pop filter (**FIGURE 9-5**). Certain parts of speech called plosives make certain letters of the alphabet come out with much greater force than others. Words that start with the letters *P*, *B*, and *T* are the most common plosives. Recorded normally, the plosive sounds will "pop" and sometimes overload the microphone. A pop filter is simply a screen placed between the microphone and the singer's mouth to stop the plosive from popping the microphone.

FIGURE 9-5

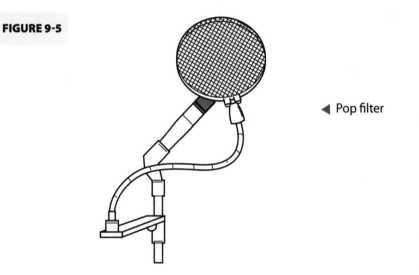

◀ Pop filter

Pop filters are cheap, and you can even make one out of wire and nylon mesh or pantyhose. An added benefit is that pop filters stop spit from hitting your expensive microphones. Gross, but it happens!

Preamplifiers

As we have touched on in Chapter 3, a microphone puts out a very small signal through its cables. In order to use a microphone with a recording device, the signal has to be amplified to be heard by the recorder. There's no way around this; every microphone needs some sort of preamplification. The good news is that most outboard mixers include a few microphone preamps, studios-in-a-box include a few as well, and so do many computer interfaces. For those devices that don't include preamplifiers, or if you've used up your available ones, you can purchase individual microphone preamplifiers (**FIGURE 9-6**). They are available in single- and multiple-channel versions. Prices start around $40.

FIGURE 9-6

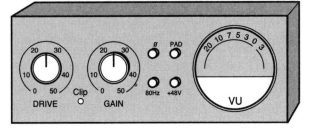

◀ Microphone preamplifier

QUESTION?

Is a microphone preamplifier considered a gain stage?
Yes, a gain stage is any device that changes the volume of a device. A microphone preamplifier boosts the weak signal up, so it is a great example of a gain stage.

Powered Microphones

Condenser microphones, except for ribbon condensers, need internal power to run, unlike dynamic microphones that induce current on their own. Condenser microphones get power either from an internal battery or, more commonly, from an external source. The power derived from an external source is called phantom power. Phantom power is delivered through the power line running from the microphone to the device it's normally hooked up to, which can be a mixer, microphone preamplifier, recording device, or computer interface. The phantom power runs through the standard, ordinary microphone cable, so no extra supplies are needed. You need phantom power only when working with condenser microphones, and most recording devices more expensive than the tape-based 4-tracks automatically give you phantom power, but be sure to check with the manufacturers if you plan on using phantom-powered microphones, just to be safe. If you already own a recording device that doesn't contain phantom power capability, don't worry; there are plenty of condenser microphones that operate off battery power, so those might be a better choice for you.

So now you've learned the basics of microphones. If you feel unsure about anything, talk to engineers and other home studio owners to find out what they use and what they recommend. Happy hunting.

Chapter 10

Mixers

No piece of gear is more central to the studio experience than the mixing board, or console as it's referred to in larger studios. This is the place where the engineer sits and does all his or her most important work. For the home studio owner, the mixer has an uncertain future because many devices have integrated mixers, and software programs provide virtual mixers that can be accessed onscreen.

Mix It Up

In the twenty-first century, the outboard mixer is no longer a necessity, although many people will always use one. In professional recording studios, outboard mixers are an essential, irreplaceable part of the studio. Even as the technology landscape continues to change, the outboard mixer in some form or another refuses to go away.

The word "mix" is defined as combining or blending into one mass or mixture. In the case of recorded audio, a mixer serves as a control over many individual sounds, or channels (tracks). A mixer has two uses: The first is to manage many inputs of sound and help to route them into the recorder, including setting the appropriate recording input level and preamplifying microphones and instruments. The mixer is also used after recording to provide balance and EQ, add additional signal effects, and provide a final master mix of your sound.

Do You Even Need a Mixer?

Twenty years ago, you had to have an outboard mixer, because there was no way around it. These days, especially for home recording, you might not need an outboard mixer. If you're using a studio-in-a-box, where the unit has a mixer section built in with effects and EQ, you can get away without an outboard mixer, because all its functions are replicated on the unit. All of the cassette tape-based multitrack recorders provide some sort of mixing control, even if it's just volume and pan controls. For computer recording, every major brand of recording software provides a virtual mixer that allows you to change the balance of audio tracks and much more, all with your mouse. With a virtual mixer, you can even record the movement of the on-screen faders to be played back automatically during the final mix down (when everything is recorded to a single file) after you've finished mixing.

If you like the idea of moving volume sliders with your hands while using a computer, you can use "control surfaces" that move the onscreen faders. You can find more about control surfaces in Chapter 17.

Benefits of Using an Outboard Mixer

If you are running a standalone digital recorder like an ADAT, DA-88, or other standalone hard-disk recorder, you won't be able to function without an outboard mixer, because those recorders provide no control of the sound during or after recording. The mixer is your link to effects, EQ, and all volume manipulation. Having an outboard mixer is an essential component in these cases.

But even if you have the ability to mix built into your equipment, you might want a separate outboard mixer. Many studios-in-a-box and computer-recording interfaces aren't exactly generous with the number of inputs and outputs they provide. Countless studios-in-a-box tout eight channels or more of simultaneous recording, yet provide only two microphone inputs. So what do you do when you want to record more than two channels of microphones? What do you do when your computer-recording device has two only inputs total? In this situation, an outboard mixer is going to come in really handy.

Whether you use an outboard mixer or a virtual mixer on your computer, you'll have access to the same features. The virtual computer emulations do just that—emulate the hardware version. Computer mixers are designed so that an engineer who knows hardware mixers will have no trouble transferring his or her skills over to a computer.

The outboard mixer will allow you to take eight microphones and mix them down to a stereo (2-track) mix that you can bring into the computer. That works great for the computer interfaces that have only two inputs. The only downside is that you lose the ability to change the volume of individual channels after you record. Whatever you set on the outboard mixer and send to the recording device or computer is what you get. That's not a bad thing if you get a good mix on the board. Some studio and live recordings are still done this way.

For recording devices that give you only two microphone channels and additional line inputs, you can take advantage of the microphone preamps that many, if not all, outboard mixers have built in. You can plug in additional microphones to the mixer, preamplify them, and send their individual outputs from the mixer to the line inputs of the recording device. This is the easiest way to increase the number of microphone channels. Not only can buying an outboard mixer be cheaper than buying a new recording device or additional microphone-preamplifiers, but having one comes in handy if you play live or have to mix sound. A few companies, such as Mackie, Behringer, Alesis, Nady, Carvin, and Soundcraft, make small outboard mixers appropriate for home studios.

Physical Layout of Mixers

One thing is for certain—mixers have lots of buttons and knobs. This can scare people at first, but like all things, if you know what you're looking at, it all makes sense. **FIGURE 10-1** shows an outboard mixer and **FIGURE 10-2** shows a screen from Pro Tools computer software. You can see that the two versions, hardware and software, look very similar in their overall layout. A mixer comprises three main elements: input/output, channel strips, and master sections.

FIGURE 10-1

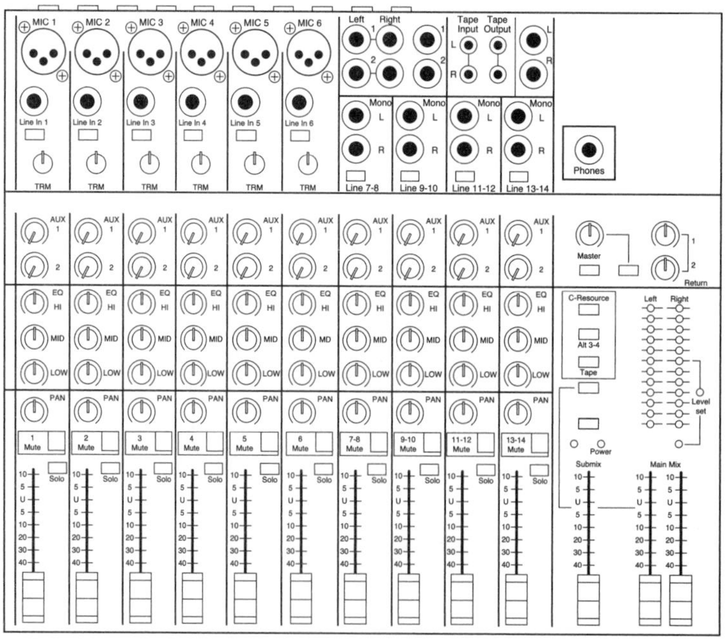

◀ Mixing board hardware

FIGURE 10-2 ▼ Pro Tools mixer software *Screenshot used by permission of Digidesign/Avid Technology, Inc.*

Input/Output

The input/output section is the basic function of a mixer. This is where the physical connections are made on the unit. For outputs, the minimum a mixer will have is a stereo "main" output, where all the individual channels get mixed to. Some mixers give you alternate "bus" outputs that you can use (more on this later in this chapter). A bus is simply a path that audio can take. In the case of mixers, it's a path inside of the mixer that the audio takes. Better mixers give you individual outputs for each channel, but these mixers tend to be expensive.

There will be one input for every channel that the mixer supports. Other inputs include channel inserts, which are effects that plug into individual channels, and auxiliary inputs, which are effects that can be assigned to any track in the mix. The better the mixer, the more aux effects you get. There is more about aux effects in our discussion on master sections later in this chapter.

Channel Strips

FIGURE 10-3 is a close-up of a channel strip from a mixer. These vertical strips are duplicated for every input channel your mixer has. If your mixer supports sixteen channels, your mixer will have sixteen channel strips, identically one after another from left to right. This takes up most of the space on a mixer. These channel strips make up most of the real estate on your mixer. They're also the reason that professional studio consoles are so massive—they contain many, many separate channels.

FIGURE 10-3

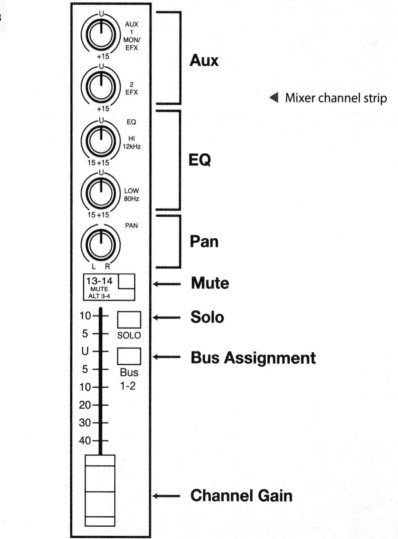

◀ Mixer channel strip

From top to bottom you might find these common elements, which are labeled in **FIGURE 10-3**.

- **Aux Section:** This is where you mix in effects that are on the auxiliary channels.
- **EQ Section:** This is where you can adjust the equalization of the sound balance between high and low frequencies of a signal.
- **Pan:** This knob adjusts where the signal is placed from left to right of the stereo image.
- **Mute:** This mutes the channel from the main mix.
- **Solo:** This mutes every other track and solos the selected track or tracks; you can use this on multiple channels to isolate a few signals together.
- **Bus Assignment:** This sends the selected channel(s) to a bus output and bus fader.
- **Channel Gain:** This is where you set the volume of each track, using either a rotating knob or a vertical slider; channel gain is used for setting volume levels after you record, not during the recording.

Master Section

So far you've learned to control individual channels, individual volume, and the like on the mixer. The master section (shown in **FIGURE 10-4**) is where you control the final output, after all the separate levels have been set. A simple master section contains one main volume control for the entire mix. You might get separate left and right channels, or the channels might be combined into one fader. A more advanced mixer goes beyond just master volume controls. If your mixer supports "buses," you will find those level controls here as well. Auxiliary effects and send and return levels can also be set here.

FIGURE 10-4

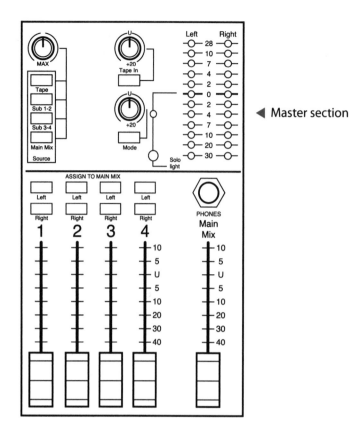

◄ Master section

Insert Effects, Aux Inputs, and Buses

An insert effect is an effect that can be plugged into one, and only one, channel. You do this via the insert jack on your mixer, which sports a tip ring sleeve (TRS) connector made for insert effects. Insert effects are typically used only for compression and EQ because you cannot mix how much of the signal gets the effect—it's all or nothing. Because compression and EQ work fine in these cases, insert effects are perfect for this. Effects are covered in detail in Chapter 15.

Aux inputs are used for effects that require a blend of effected and noneffected signals. Effects like reverb, chorus, and delay sound terrible when you hear only 100 percent effects. Aux effects are also used when you wish to use the same effect on more than one channel. An aux input is plugged into the master section aux input/output, and each channel has a control for how much of the effected signal to blend in with the dry, unaffected signal. This is great when you have hardware effects processors and you have to make the most out of a few pieces of gear. On the computer side, aux effects do the same as their hardware counterparts. You can usually have as many aux effects as you want in software, because you can reuse the same plug-in on multiple tracks.

Buses act like a submaster fader. That is, you can send a bunch of channels to a bus, and the bus fader will raise or lower all of the signals fed into it together. For example, let's say you're mixing a drum kit. You have four channels of drum sounds and you have the perfect mix between the individual drums. By itself, the drums sound great, but when you add in other instruments, you notice the drums' volumes are a little low. You could raise each drum track one by one, but that means that you lose that perfect balance you worked so hard to get. If your outboard mixer or virtual mixer supports buses, you can assign all the drum tracks to bus one, which sends all four signals to one volume knob (the bus). Then when you turn the bus up, all four signals fed into the bus get louder while retaining their individual balances. For outboard mixers, multiple buses are something you gain with the better mixers. The more you spend, the more buses you get. Just about all the major computer recording software supports buses. It's a great tool for recording, and is commonly used to balance groups of instruments like drums and double-miked instruments.

Bus and buses are also referred to as "sub" (for subgroups) and "subgroup outputs" on many mixers. They mean the same thing.

Buses are also used when interfacing with standalone recorders like old-style tape machines and the hard-disk and digital tape recorders available today. Many of those units allow only eight inputs total, so buses comes in handy if you need to consolidate four drum microphones to one or two channels, again regaining control of the final mix, not just the individual channels.

Getting Sound into a Mixer

Getting sound into and out of a mixer is really important. It's easy to make mistakes here. The first place to start is the microphone channels. Attach your microphones to the microphone channels on the board. If your microphones are condenser microphones that require phantom power, and your mixer supports it, flip the switch on the top or side of the unit that supplies phantom power. Now your microphones will be powered up and ready to go. Next, connect your line-level interfaces, such as your keyboards and line outputs from amplifiers and direct boxes.

Setting Recording-Friendly Levels

For each channel you record through the mixer, you need to set the level of each input. Activate the solo button on the channel you are working on to isolate the sound. For microphones, set the input trim control so the microphone's loudest sounds do not clip the input. If your mixer has a meter that lights up (most do), make sure the loudest sound stays in the green lights and doesn't hit the red. Repeat this step for each microphone you have.

ALERT!

When setting microphone levels for recording a drummer, make sure the drummer hits the drum really loud so you can get an idea of when clipping might occur. Set your gain so that the loudest hit does not clip.

Line-level instruments connect to the line channels. On the keyboard or guitar amplifier, you should adjust the gain on the amplifier, not on the recorder. The fader for those tracks should be set to "unity gain," which means nothing added, nothing subtracted. Unity gain is marked with a zero. This allows you room in either direction for volume changes later on. Also, check the level meter to make sure you aren't clipping. Taking time like this will ensure that you don't clip and distort the inputs while you're recording.

Measuring Electrical Voltage

"Decibel" is a confusing term. Scientifically, the decibel is not a concrete measure of any one thing; it's a ratio of power or intensity to other factors.

There are actually a bunch of different decibels that we deal with. The basic decibel measures sound pressure. This measures the loudness of the sound pressure created. For instance, the sound of a plane taking off is about 120dB.

On a mixer, decibels are used in a much different way. The first way is as a measure of electrical voltage; the second way is as a level of sound output. When you look at a mixer's slider, you will see that it's marked up and down with decibel marks, ranging from negative decibels to positive decibels. See **FIGURE 10-5**.

FIGURE 10-5

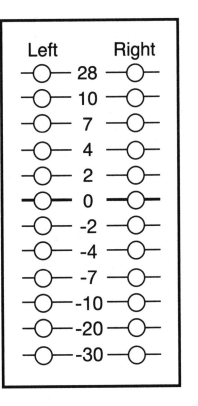

◀ Mixer levels

Negative decibels? This seems to go against what we know of decibels. If you're comparing a volume slider to sound pressure, you will be confused. Think of the volume slider simply as a way of boosting or lowering the prerecorded signal. Negative dB values mean you are reducing the level that's been recorded, and positive dB values mean that you are raising the recorded output.

So, to sum this up, as you mix, place all of your volume sliders at zero. At that setting the mixer is not artificially boosting or cutting anything. You hear the signal that you recorded from the instrument or microphone itself.

Output Scale

More decibel madness! When we discuss using faders to boost and cut input volume, we might naturally push the vocals +2dB above normal to make them louder. Here's where the other side of the decibel confusion erupts, because when it comes to final output, decibels are measured using an entirely different system. We're now faced with a system called dBFS (decibel full scale). Decibel full scale simply considers 0dB as the absolute *loudest* signal you can have. You cannot pass 0dBFS or clipping will occur, which many of you have experienced. Since decibels aren't a fixed ratio, dBFS calls 0 the loudest and works backward. When mixing or mastering music, especially in the digital world, you can never exceed 0dBFS, ever. If you do, you will distort and clip the signal.

Maximize Music; Minimize Noise

We've talked about setting levels high enough so that they are full and don't clip. The reason for this is that the louder the input signal, the less noise the signal will contain. Even if you are planning on using a soft track, record the input high. Turning the track down later will help cover up any noise in the signal. Volume faders on mixers or computers can boost only a few decibels up. They can, however, reduce signals to nothing. It's better to have too much signal and reduce it later than to have too little signal and try to overboost.

FACT

Every signal, no matter how well recorded, will have some noise in the signal. This isn't usually a problem if you set the correct signal-to-noise ratio by recording tracks properly and setting the correct input levels. If the signal is hot enough, the noise won't be an issue if you've paid attention to your settings.

Interfacing with the Recording Device

How you interface with your recording device depends on how elaborate your mixer is. For users with a simple mixer, the mixer outputs to separate left and right channels. This means you have two tracks to send to the recorder and two separate signals to use: right and left channels. Using the pan controls on the input channels, you can send certain sounds to the right cable and certain sounds to the left cable. Using a mixer like this doesn't mean you lose total control of the sounds. By panning certain sounds all the way left or right, you force that signal into one of the two cables. Each of the output cables can go to a separate track on your recorder. If you have buses on your mixer, those buses will also have outputs, also stereo left and right. For every bus you have, there will be another two outputs to your recorder. A decent-quality four-bus mixer will give you a total of four extra channels of output to your recorder. If you pan correctly and set the bus assignments well, you can route tons of signals flexibly to your recorder. If you use a lot of microphones and your recorder doesn't support many microphone inputs, this might be a cost-effective way to go—it's cheaper than buying all those microphone preamps separately.

Chapter 11

Recording Individual Instruments

No matter when you sit down to record, there will be different variables in the recording process. The type of song you are recording and the instrumentation will change from time to time; your approach will also change along with it. It's time to get your hands dirty with some specifics of instruments and how to record them.

Guitar and Bass

Guitar can be one of the easiest instruments to record—and other times the most difficult. Back in the early days, musicians stuck a microphone in front of the amplifier and that was it. Now we are barraged with different amplifiers, effects processors, amplifier simulators, and now realistic plug-ins in the computer-recording arena.

Direct Interfaces for Guitar and Bass

To record guitar and bass without the use of an amplifier, you will need a direct box (shown in **FIGURE 11-1**) in order to properly interface with the mixer or recording device. Some studios-in-a-box and computer interfaces feature a Hi-Z (high impedance) direct guitar/bass input. If you have this feature, you won't need a direct interface to record your instrument—you can plug right in. These are becoming more and more common, especially with studio-in-a-box setups.

FIGURE 11-1

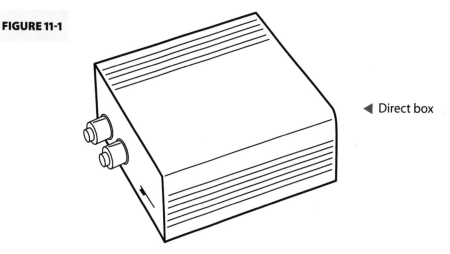

◀ Direct box

Direct instrument signals are usually high-impedance sources, so they can't be plugged into a mixer or recording device. In order to record them, you need to use a direct box to convert the high-impedance, unbalanced signal coming straight from the instrument to the low-impedance, balanced output. The output of the direct box will be a balanced cable, either XLR microphone or tip ring sleeve (TRS).

If your amplifier is equipped with a line out, you can run a cable directly from the back of your amplifier, directly into your mixer or recording device's line input. The amplifier line out will have the correct output level and impedance. The amplifier's line-out jack is specifically designed for studio setups like this. You can also use this feature for live sound applications.

Unfortunately there's one catch to plugging a line out of an amplifier—you might encounter a ground loop, which can cause some very unpleasant buzz in your signal. If this happens, plug your instrument into a direct box that has a ground lift switch. The direct box will take care of the buzz and you will have a good clean signal to work with.

FACT

Using a direct output from an amplifier doesn't defeat the speaker; it will usually still play unless your amplifier has a special "silent record" feature. The direct output is useful when microphones and microphone inputs are scarce.

Miking an Amplifier for Guitar and Bass

Miking a guitar or bass amplifier is pretty simple. Most engineers simply place a dynamic microphone a few inches from the center of the speaker. Placing a microphone close to a sound source is commonly called "close-miking." Close-miking is commonly used with amplifiers, some acoustic instruments, and most drums, but not drum overheads.

The exact location of the microphone on the speaker will differ from amplifier to amplifier and microphone to microphone. It's common to place the microphone slightly off the center of the speaker. You can get different tones based on the location of the microphone. As you move the microphone farther to the outside of the speaker, the sound warms up. The closer you move to the center, the more grit and high end you can capture. Let your ear be the judge of what sounds best. Be prepared to spend a good amount of time moving the microphone around at first, until you learn what works for you.

FIGURE 11-2

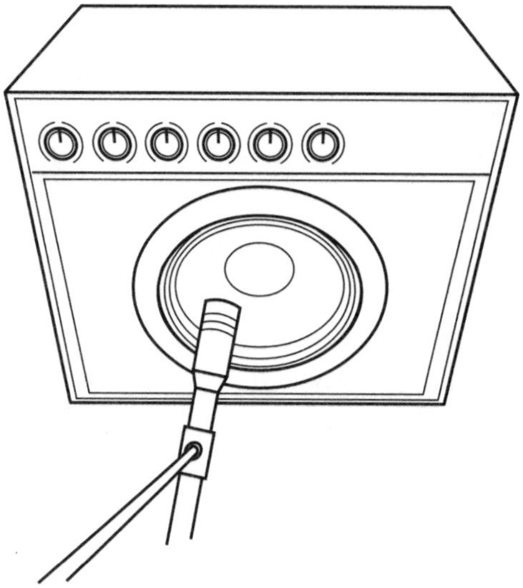

◀ Miked guitar amp

Recording engineers employ a few tricks to place a microphone on an amp that might help you achieve some unique sounds. First, you don't have to use a dynamic microphone. Many condenser microphones can handle high sound pressure level (although not the very highest), so if you're not cranking the amplifier to 11 (as in the movie *This Is Spinal Tap*), a condenser could be fine. You might get a richer sound by using the full range provided by a condenser.

Another trick is to use two microphones: a dynamic microphone to close-mike, and a condenser microphone farther back to catch ambience. By blending the two signals together, you can get a nice, rich sound. The farther away the condenser microphone is, the more ambience and natural reverb you'll get in the sound. Many engineers employ this technique to fatten up and widen their guitar and bass tracks.

Direct-Recording Preamplifiers

There's a new wave of digital technology available to guitar and bass players: the direct-recording preamplifier. Devices like the Line6 POD and the Behringer V-AMP emulate amplifiers and effects in one handy unit. What's even better is that they can output to line level, allowing you to plug them in directly, bypassing the need for an amplifier or direct box. These units are small and compact, and have become a staple for guitar players who record frequently. Many of the guitar sounds you hear on TV, radio, and studio recordings might very well have come from direct-recording pre-amplifiers.

Wet or Dry Effects?

Effects such as reverb, delay, and chorus are part of the unique tones that make modern guitars and bass sound the way they do. Most players come into the studio with a "sound" they always play with. Typically this sound is achieved with added reverb and other effects. The big question is whether or not to record the guitar with effects (wet) or without effects (dry). Some players consider the effects a signature part of their sound; it would be hard to duplicate those sounds later on when mixing. However, certain tones sound very good by themselves—but how they fit into a full mix is another story. The biggest disadvantage to using wet effects is that you have no control over the effects after they're recorded. If you suddenly find out during mixing that there's too much reverb and the guitar sounds distant, there's little you can do to fix it. It's usually safer to leave reverb out of guitar and bass sounds, because it's easy to add it during mixing. That way you have control over the final sound.

Computer Plug-Ins for Guitar and Bass

Computers have become a mainstay of recording studios. Both at home and in the professional world, plug-ins for DAW (digital audio workstation) programs are developing at a dramatic rate. Guitar and bass players have not been left out of this party. Companies such as Line6 and Johnston Amplification use digital modeling in their guitar amplifiers using DSP (digital signal processing). DSP is how computers process recorded audio for effects. One of the coolest plug-ins available for guitar and bass is AmpliTube by IK Multimedia. Native Instruments now offers Guitar Rig, an impressive guitar amp and effects modeler as a plug-in.

As you can see from the screenshot in **FIGURE 11-3**, the controls mimic what you're used to seeing on amplifiers. Not only are there a variety of modeled amplifiers, the effects are built in, too! You simply plug your guitar directly into the computer and AmpliTube does the rest.

One of the ultimate features of a plug-in like this is its ability to change the sounds after you have recorded them. The audio track contains only the clean audio; it's the plug-in that does the rest of the work. That means that you can change the sounds as you go along, without having to rerecord the part. That's something that wasn't possible before these programs changed the way we work.

Isolating Amplifiers

Many guitar amplifiers, especially tube amplifiers, sound best when turned up quite a bit—or even turned *all* the way up—because of how tube amplifiers saturate and create gain. To record with the amplifiers this loud, isolation of some kind is a must. Whether you banish the amplifier to a closet, another room, or a special isolation box, for the sake of your hearing, you need to do something. This is especially true when recording loud amplifiers with a band. Isolation can minimize the amount of bleed-through into the other instruments' microphones.

Vocals and the Spoken Word

These days, getting good vocals or spoken words on tape or hard disk can be one of the greatest challenges of recording. The human voice is one of the hardest sounds to re-create because listeners are supercritical if it doesn't sound exactly as it should. We are, quite naturally, very used to hearing the human voice.

Choosing the Right Microphone for Voice

If you ever had the occasion to sing on stage, you probably sang into a dynamic cardioid microphone. Dynamic microphones are great because they handle loud sounds well, reject other noises from the sides, and are virtually indestructible. While dynamic microphones might be perfect for the stage, they aren't always perfect for the studio. Dynamic microphones don't re-create the full frequency spectrum very well, which can be a problem that sometimes shows itself in unnatural sounding vocals. However, a dynamic microphone is your best choice if you work with a loud singer who screams his or her head off (or you are one yourself), because it handles the sound better. For everything else, choose a large-diaphragm condenser microphone. Large-diaphragm condensers are those microphones with the large, fat heads. They allow you to pick up all of the frequencies present in a human voice.

FIGURE 11-4

◀ Vocal miking

Save your best condenser microphone for your vocal work because it will allow you to pick up the nuance and detail of the human voice. If you work with vocals a lot, investing in a good microphone will pay off time and time again.

Pop Filters

When we sing and speak, we create certain sounds, called plosives, that tend to "pop" the microphone with sudden bursts of air. Words that begin with the letter *P* are the most notorious examples of plosives. Plosives are bad because they can kill your recording by distorting or clipping the signal either to the microphone or input channel. Plus, they sound just plain awful. On digital recorders, anything that is clipped is instantly turned into garbage noise—a square wave, to be exact. Clipping always ruins whatever you are recording. Lucky for you, there's a simple solution—the pop filter.

A pop filter is a small round disk of mesh that sits between your mouth and the diaphragm of the microphone. When the plosive leaves your mouth, the excess air is filtered through the mesh, and only the normal sound reaches the microphone. Pop filters are a must for vocal recording and can be obtained at a relatively low cost. You can even make one out of a coat hanger and pantyhose (new ones, please!). Pop filters also serve another important purpose: They keep saliva from reaching the microphone.

The Proximity Effect

Microphones exhibit a special phenomenon called the "proximity effect." Simply, the closer you stand to a microphone, the more bass frequencies come through. As you step back, the bass diminishes. If you are close-miking a vocalist, this phenomenon creates a muddy sound. The easiest option is to have the singer step back a little, but then the track might sound distant. If so, you can do a few things to fix the problem. Many condenser microphones are equipped with a bass roll-off switch. Turning on the base roll-off switch will usually cut the bass frequencies from 200Hz and below, which is where most of the proximity effect occurs. If your microphone doesn't have such a switch, you can EQ either on your outboard or virtual mixer by cutting 200Hz and below.

FIGURE 11-5 shows what a 200Hz roll-off looks like in Logic's EQ plug-in.

FIGURE 11-5 ▼ Logic's EQ plug-in *Screenshot used by permission of Emagic Soft and Hardware GmbH/Apple Computer.*

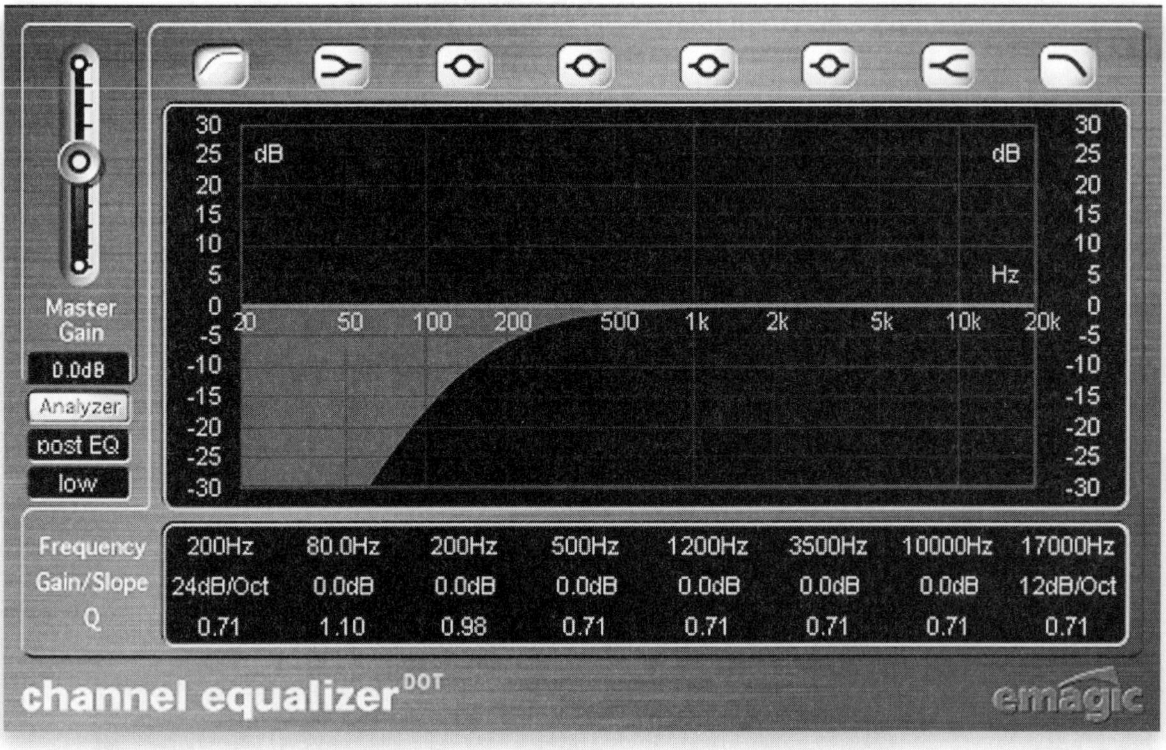

Frequency	200Hz	80.0Hz	200Hz	500Hz	1200Hz	3500Hz	10000Hz	17000Hz
Gain/Slope	24dB/Oct	0.0dB	0.0dB	0.0dB	0.0dB	0.0dB	0.0dB	12dB/Oct
Q	0.71	1.10	0.98	0.71	0.71	0.71	0.71	0.71

Voice Monitoring

The vocalist should monitor through closed-ear headphones. The level of the sound coming to the vocalist's ears is critical to getting the best performance. Hearing too much or too little of his or her own voice can cause major pitch problems. Adjust the mix until the singer feels comfortable and you notice that the pitch remains relatively constant. Typically, a singer doesn't enjoy hearing the dry sound of his or her voice through the headphone mix. Reverb is usually needed to sweeten up the sound and make the singer more comfortable.

Isolating the Voice

Vocals are something you definitely want to isolate, because it's important to cut down a lot of the excess ambient noise in the room. For the home studio, that could mean sending the vocalist to the bathroom to sing! Isolation will give the voice a relatively flat, neutral sound, which offers you the most control over the sound. Ambience can be added with reverb later.

Holding Pitch

For a singer, there are few things more important than holding good pitch. Not everyone is blessed with this ability, so you might have to rerecord time after time until you get something you like. But that can lead to frustration on the part of the singer, which can make the situation worse.

FIGURE 11-6   ▼ Antares Auto-Tune   *Screenshot used by permission of Antares Audio Technologies.*

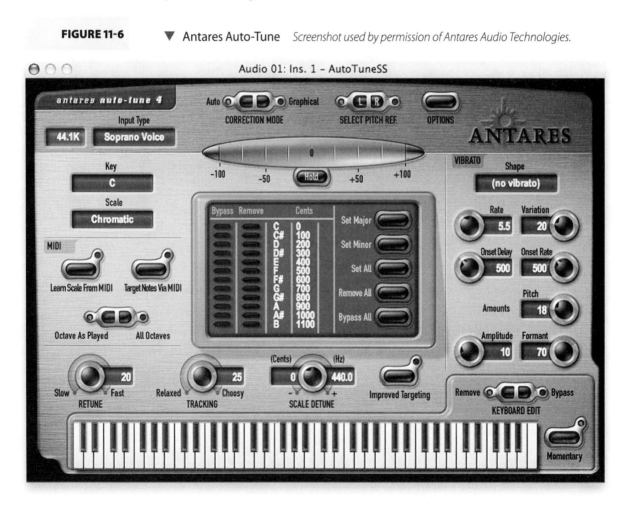

Technology to the rescue! Antares Audio Technology invented a product called Auto-Tune that has changed the way vocals are recorded. Originally conceived as a computer plug-in, Auto-Tune has taken the audio world by storm. Auto-Tune (shown in **FIGURE 11-6**), which is available either as a stand-alone rack unit or a computer plug-in, listens to the vocal track and corrects the pitch of the notes as they stream by. Not only does it correct the notes in real time, it also does it invisibly—you can't tell that it's there . . . except of course that a perfectly in-tune vocal is coming out the other side. It also works for other instruments besides voice! Single-line instruments only please!

Most vocalists sing a touch flat or sharp on certain notes, and those small differences are easily corrected by this amazing plug-in. However, no piece of technology can change extreme mistakes in pitch and singing. The more pitches that Auto-Tune has to pull up or down, the less transparent the changes will be. If you work with singers often and pitch is important to you, plug-ins like this might come in very handy.

Electronic Keyboards

Good news! The keyboard is one of the simpler instruments to record. Any keyboard worth its salt is equipped with line outputs. The line out plugs directly into your mixer or recording device. Better yet, you won't have to eat up one of your valuable microphone channels, because the output level is sufficient for recording (that's what line level means). You can set the level on the keyboard simply by adjusting the overall output level.

Recording in Stereo

Keyboards, especially modern synthesizers, utilize a stereo signal path internally. That means that most of the patches are designed for listening in two-channel stereo—left and right signals. If you output only one of the signals to your recorder (perhaps to save inputs), you might not get a full sound. This is due to how the keyboard pans the sound from side to side. To get the best sound, use two tracks for a keyboard. The exception is, if you're trying to record an older analog-style keyboard that you know has a mono output, then you can get away with one track. It's also possible, through programming your synth's internal parameters to "sum" the output to just one channel, to combine the two signals into one.

FACT

The term "sum" means to combine many signals. A mixer is a great example of a device that sums many signals into one stereo pair.

Monitoring Yourself

For monitoring, you have many options. If you're laying down solo keyboard tracks, you can just plug a headphone into the keyboard. If you play with an amplifier, you can plug into that as well. If you're working with prerecorded tracks, you can monitor from the recorder with either headphones or monitor speakers.

MIDI or Audio?

Keyboards communicate with computers via MIDI. One of the benefits of using MIDI, simply, is that it's *not* audio, meaning you've got a lot more ability to edit the sound. With its simple code of commands to the keyboard, MIDI can be highly edited on a computer screen through all the major DAW programs. On the other had, if you record the keyboard signal as audio through an audio track, there's little you can do to change aspects of the performance. You would have to go back through the piece, decide what you want to change, and rerecord the audio. If you keep the performance as MIDI, you can tweak the part endlessly until you feel it is final.

At some point, however, you need to record the keyboard's actual physical sound. By itself, MIDI and MIDI sequencers don't capture any "audio" that you can print to a CD. At some point you have to plug the audio outputs from the keyboard into a recorder. Recording the audio from the keyboard allows you to make the final mix, and bounce a final product to an audio CD.

ALERT!

Many keyboards contain built-in sequencers, which are recorders for MIDI. While the built-in sequencers allow you to record and make music with the keyboard's built-in sounds, the screens are small and editing can be cumbersome. This is why computer-editing programs are so popular with keyboard players the world over—big screens and mice make for powerful editing.

Drum Sets

How to best record a drum set is a slightly complicated subject. Most of the difficulty lies with how many sounds you have to capture at once. A basic drum kit consists of:

- One snare drum
- Two toms (possibly three if there's a floor tom)
- One bass drum
- Two overhead cymbals
- One hi-hat cymbal

For those of you who weren't keeping score, that's up to *eight* different sounds to capture. And that's just for a basic kit; you might encounter far more complicated kits when you record.

How to Mike a Drum Set

For miking a drum set, there are a few different schools of thought on the subject. The first school says "more is more": Place a microphone on as many separate parts of the drum set as you can so you can control the level and balance as you mix. Here is an example of a fairly extensive microphone setup for a basic drum set:

- One dynamic microphone for the snare drum
- Two dynamic microphones for the two toms
- One dynamic microphone for the bass drum

- Two overhead condenser microphones to pick up the two overhead cymbals and ambience of the drum set
- One dynamic microphone for the hi-hat cymbal

That's a total of *seven* microphones! That also means eating up *seven* microphone channels on your recorder. You could pair down (or sum) the inputs to fewer channels, but that would defeat the purpose of miking everything. If you have enough inputs, and you can spare doing it this way, there's nothing wrong with it. You will have great control over the sounds when you mix. Most professional studios use multiple drum channels. However, this might not suit your needs well when you consider everything else.

The other school of thought says "less is more": Use fewer microphones for a more ambient sound. The thought behind this is that a good drummer takes care of his or her own balance, so there *should* be little need for extensive tweaking. This is not always true; it really depends on the drummer. At a minimum, you can get away with two overhead condenser microphones to pick up the whole drum set. Although you can get a nice sound that way, it might not be flexible enough. Many engineers like to put effects on parts of the drum set, and none on others. Snare usually gets some reverb, for instance. A simple overhead-miking setup won't give you the ability to add effects. For the most flexibility, the minimum you really should use for the basic drum set is four microphones, as follows (shown in **FIGURE 11-7**):

- Two overhead condenser microphones
- One dynamic microphone on the bass drum
- One dynamic microphone on the snare drum

Using that setup will give you a more open sound that you won't have to spend days mixing together. Putting the separate microphone on the snare drum allows you to tweak its sound and add reverb if necessary. Putting the separate microphone on the bass drum allows you to set the EQ on that drum if it gets too much bass and sounds "muddy." You can get a great drum sound this way.

FIGURE 11-7

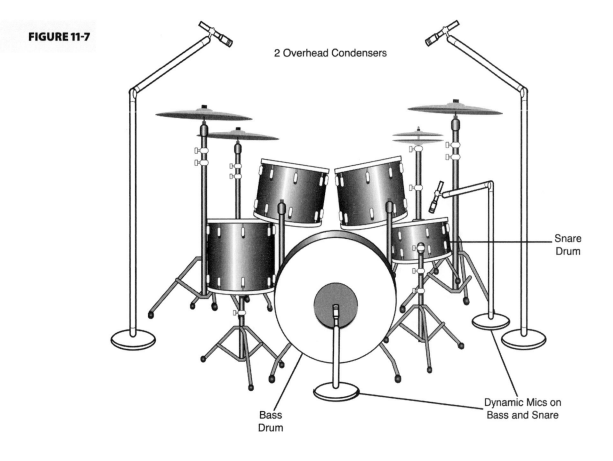

2 Overhead Condensers

Snare
Drum

Bass
Drum

Dynamic Mics on
Bass and Snare

▲ Drum-set microphone placement

Muffle

Bass drums, also commonly known as "kick" drums, are the lowest frequency in the drum set. Because the drum shell is so large, it's really easy to get an overly "thumpy" bass drum sound, no matter how hard you try to reposition the microphone to get rid of it. You can easily solve this by placing a muffle inside the bass drum to get rid of some the excess ring and thump. Towels, pillows and blankets work well in this regard. Experiment with the sound to see what you like. Some bass drums sound fine without any help, so only when you test the microphones, will you hear what needs to be done.

Tune Up!

Tune your drumheads. Better yet, replace them with new ones, and then tune them before a recording. New drum heads sound great and record well. There are many books available on this subject, just check out your local bookstore or music shop for information on proper drum tuning. Having a properly tuned drum set can make all the difference in the world.

Acoustic Instruments

Acoustic instruments can be more difficult to record well. You'll need to practice a few times until you get a sound you're happy with.

Winds and Brass

The wind instrument family can be difficult to record because of the very nature of how the instruments produce sound. First, point the microphone where the sound comes out. The difficulty with a lot of the wind instruments is that the sound actually comes out of the whole instrument—finger holes leak sound and so on. So where do you microphone? This might seem simple . . . but certain instruments might fool you.

To cut down on breath noise from a flute, try using a pop filter in front of the microphone. The filter will cut the air from reaching the microphone and sounding overly "breathy."

Trumpet and the other brass instruments have a very clear bell where the sound shoots out. Simply place a condenser microphone slightly in front of the bell and you're good to go. For loud brass players, place the microphone back slightly to ease possible distortion. Flute is a more difficult one. While it might seem as if the sound comes out of the end of the instrument, most of the sound actually comes out from the head joint, where the player blows. The problem with this is, that's also where all of the breath noise is, which can make it very difficult to get a clean flute sound. Place the microphone just slightly next to the flute's head joint to pick up a good combination of some body sound, head joint sound, and minimal breath noise. Use a large-diaphragm condenser microphone that uses either an omnidirection or cardioid pattern.

Instruments may sound good by themselves, but when you add them to the rest of the mix, they may not fit as well. This is a common problem and can usually be fixed with some creative EQ and effects placements.

In saxophones, the bell delivers most of the sound, but there is also sound from the keys as well. Try to position the microphone in a way that you can clearly hear both sounds; then move the microphone around till you get a sound you like.

Piano

Acoustic piano is a difficult sound to get just right. If you are lucky enough to have a grand piano, open the lid to expose the strings. The best way to record piano is to use two condenser microphones, one devoted to the top strings and one devoted to the bottom strings. You can pan one microphone hard left and the other hard right to get a fairly wide-sounding stereo image (see **FIGURE 11-8**).

FIGURE 11-8

◀ Miking a piano

If you use just one microphone, choose an omnidirectional microphone for this job and place it dead center to get the best sound you can. For those who use upright pianos, which are far more common, take off the top of the piano's case to expose the strings. Follow the same technique noted earlier for either a single or double microphone setup.

Strings

For any single string instrument, whether it's a violin, cello, bass, mandolin, guitar, or anything else, place a condenser microphone close to the sound hole or *f* hole to get the most direct sound. You can place the microphone fairly close to get a good strong signal without much fear of distorting the microphone. **FIGURE 11-9** shows an example of an acoustic guitar miked.

FIGURE 11-9

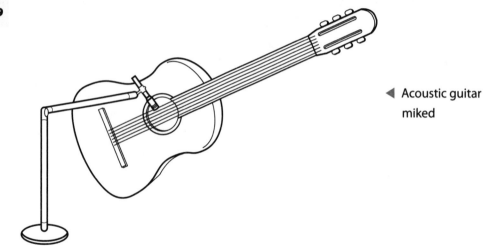

◀ Acoustic guitar miked

On bowed instruments, you might get excess bow noise if you are too close, so move the microphone around until you get a rich, pure sound. For large groups of strings, like string quartets, or groups of guitars, you can either close-mike each individual instrument, or use one or two condenser microphones set back to get the sound of the whole group. Stereo microphones work well for this sort of thing too.

ALERT!

The biggest problem recording a large group is microphone bleed. Microphone bleed occurs when a microphone unintentionally picks up the sound of other instruments nearby. To minimize this, try to isolate your sources with baffles and or room dividers, and use cardioid microphones.

Stereo Recording

When you record acoustic instruments, or anytime you want to capture a stereo recording, you will utilize something called the "XY" technique.

FIGURE 11-10

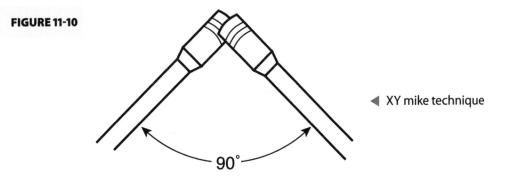

◀ XY mike technique

90°

The XY technique uses two microphones crossing at their heads, pointed in opposite directions, usually at a 90-degree angle from each other (see **FIGURE 11-10**). This works great on acoustic instruments and any time you want to record a simple stereo mix.

Chapter 12

Editing

W ord processors redefined the publishing industry with three simple words—cut, copy, and paste—and a new form of editing was heralded in. Musicians got their first taste of the power of a similar kind of editing with MIDI in the 1980s. Digital audio editing is now available to just about everyone. Sorry, cassette-tape multitrackers . . . the digital folks get to have a lot more fun.

Making Changes

Editing is the art of altering a performance. While it might seem counterintuitive and nonmusical to go back and change things, the reality is that no one is perfect every time. The outtakes from movies come to mind. Those veteran actors are sometimes unable to keep a straight face or deliver the correct line. It's also very common for a motion picture to be shot out of sequence and added in later. What you see at the end looks cohesive, but it might not have been shot that way. Music is no different. It's possible to record separate instruments, heavily edit the performance, move sections around, and have it sound perfect—like one straight take.

Some styles of music rely on edits less than others. Classical music is almost never edited. Jazz music is often not edited either, although many artists edit some parts—just not the solos, which are typically left intact. Rock and pop music might be highly edited; in fact, that's usually the case. If your band or project can get it right the first time, more power to you. For everyone else, welcome to editing!

What Editing Used to Be

Back in the good old days of analog recording, editing involved physically cutting the tape and splicing a new section of tape. Edits were done on a special block of metal called a splicing block and the cuts were preformed with a razor blade . . . ouch! Editing like that was difficult to say the least. Finding the exact spot to make the edit and getting the new material to line up perfectly was no small feat. With multitrack tape, everything got more complicated. Each track occupied a small section of the tape. Making an edit to just one track meant cutting a small window in the tape and pasting in a new one. Edits were used to fix only blaring mistakes and other tragedies. Many engineers would push for a better take rather than perform miracle surgery. Editing was seen as a last resort.

What It Is Now

Editing sure has changed! Digital audio has changed the way we all work, and it's one of the main reasons the home studio is so powerful—we get to edit just like the big boys! Digital audio recording, whether it's done via a studio-in-a-box or a computer DAW (digital audio workstation) is based on the principal of nonlinear editing. Tracks don't have to be in line together as they are on a tape. A digital audio mix is simply several audio files read at the same time off the disk; they can reside anywhere on the hard disk. This is because the files don't need to be read by the recording machine or computer as if they were sentences. Because of this, they can be edited with great ease.

For example, suppose you're recording your latest hit. You are laying down the guitar track, and you mess up the melody in the second half of the song. Coincidentally, you played the *same* melody earlier in the track and you played it perfectly. Do you have to scrap the whole track? Maybe years ago you would have, but not now. Just copy the correct performance from the beginning of the song and move it to the end. Intrigued? Read on.

Just Because You Can . . .

Just because you have the tools to edit with an amazing degree of accuracy in the digital world does not mean you should. Let's use Band X for an example. You buy Band X's album and you think it sounds great. The CD is well produced, all the instruments sound great, the performance is top-notch—in short, this band is kicking. You purchase concert tickets eagerly anticipating seeing Band X live. The day comes and you head out to the concert. The lights, the stage . . . Band X takes the stage and sounds horrid. The signer can't sing on pitch at all, and the band is falling apart. You leave the concert very upset. What happened? Did Band X have a bad night? Maybe so, but more likely they might have fallen into the trap of "overproduction." That is, the band might have recorded their music one track at a time and perfected each track before releasing their CD.

ALERT!

There are plenty of situations where a simple stereo (left and right) recording will more than suffice. Multitrack recording can put artificial control over a group's sound, balance, and vibe. Before you dive with both feet into the multitrack arena, try a simple stereo microphone setup and see what you think. You might save a lot of time and trouble in the editing department.

Editing can be a bit of a trap, especially the high-precision computer editing where you can change a single note of a solo you thought was "off." The result might be a standard you can never replicate live. You want your work to be perfect, and you should try to make it as good as possible, but it's so easy to get carried away. Just because you have these tools doesn't mean you should overuse them. Fact is, a great album isn't a perfect album.

Tape-Based Editing

Owners of cassette tape 4-tracks can edit, too, but the reality is their options are very limited. The cassette tape was heralded as a durable medium for music in reaction to easily ruined vinyl records. The tape is concealed in a hard plastic coating, which makes it much more difficult to damage, but also *very* difficult to edit, if not impossible. While it's possible to open up a cassette tape and expose the raw tape, it's hard to put it back together. You're also faced with the challenge of finding the spot on the tape you need to cut. Unlike open-reel analog tape machines that let you see the tape pass the playback head and allow you to advance the tape by hand to find the correct spot, most cassette multitracks are covered and don't allow you to really see what's going on. This makes finding the exact spot almost impossible, so many musicians just leave well enough alone.

If you're the adventurous type, you can open the tape, make edits using the correct splicing tape, and put everything back together—if you're lucky. You could also dump tracks to other tape recorders and rerecord them in sections, which is also very difficult. But if you're going to go to this much trouble, maybe it's time to think about upgrading.

Computer-Based Editing

Since computers rely on nonlinear editing technology, DAW programs give you a great deal of freedom in moving around information. Your masterpiece of music is nothing more than chunks of data to a computer. It doesn't care what order they're played back. By default, the computer plays back your music just as you recorded it. What's different is that you can drag pieces of your music to and fro at will. Let's take a look at a sample Pro Tools session for some visual aids.

FIGURE 12-1 ▼ Pro Tools session *Screenshot used by permission of Digidesign/Avid Technology, Inc.*

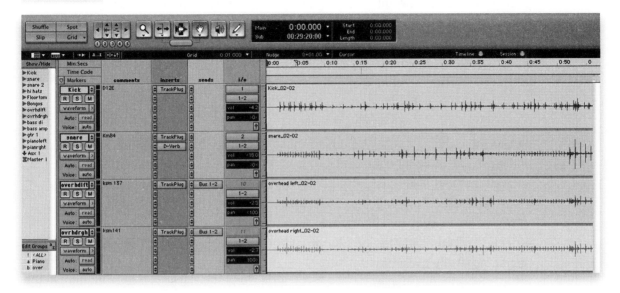

FIGURE 12-2 ▼ Edited Pro Tools session *Screenshot used by permission of Digidesign/Avid Technology, Inc.*

As you can see from the simple 4-track mix in **FIGURE 12-1**, each track occupies its own line. The different tracks are shown by visualizations of what the audio looks like. Suppose you wanted to add space at the beginning for an intro that you were going to write and record later. At the same time, you want to double the length of each track by copying them end to end. Take a look at the finished product, shown in **FIGURE 12-2**.

You'll see that the window is a bit smaller to accommodate the length—that's purely so you can see it all in one window. All we did was drag all the files to the right by thirty seconds to make room for the intro. After that we made a copy of each track and put it end to end with the original, which allowed us to repeat a part of this song to double its length. Simply done. All it took was some dragging with the mouse.

As you can see, it's very simple to build up arrangements and change aspects. And this is just the tip of a very large iceberg, because this is only the beginning of what you can do.

Editing on a Studio-in-a-Box

The standalone studio-in-a-box handles the editing of audio data a bit differently than a computer does. While many standalone systems have displays to show you track information, very rarely can you see great detail. There are also no mouse and keyboard to help you. But this doesn't mean you can't do some amazing editing. These systems have fully operating cut, copy, and paste functions. The difference is that instead of working with a visual representation of the music, you specify exact timing in a text-based editing menu. It wants to know what track and what time in minutes, seconds, and milliseconds to start the edit; when to end it; and then where to put it. Actually, many folks who got their start on these systems and went to a computer later found the mouse too inexact for them. You get to specify exactly where to place the files, with great detail. Studio-in-a-box systems allow you to do very precise editing this way. It's also possible to build up arrangements in the same way the DAWs do. Using time-based editing, it's possible to move the section at :30 to later in the track, or anywhere you like. All you need to know is where to start and stop the edit, and where you'd like to move it.

FACT

"Oops, I liked the old version better!" A great feature of the DAWs and studio-in-a-box systems is the undo command. You can revert to an older version with ease. All the editing is done "nondestructively" so you can back up and change your mind anytime.

Digital Micro-Editing

Editing is regularly used to fix mistakes and redo parts that weren't quite up to par. But editing can also take you in creative directions you never thought of. Sometimes creating arrangements from disparate sections can yield some really exciting results.

We've all sung a few "special" notes in our day, notes that just stand out and say "I'm out of tune!" With micro-editing it's possible to change just one note in a phrase. It's hard to do because two things have to be going your way for this to work. First, you have to be able to isolate that one note, which might be a chore in itself. Second, the new edit that you add in has to sound natural, not as if it was added in later.

Now in order in order to micro-edit, whether you're taking out one note or a whole solo, a few things need to happen. The first is zero crossing.

Zero Crossing

Simply put, sound is a combination of frequency (the pitch of a sound) and amplitude (loudness). To edit well, you need to find what are known as zero crossings. A zero crossing is a part of the audio where the amplitude or volume is zero, which happens quite often. To find this, you need to zoom in on the waveform on the computer or use the "find zero crossing" function on the studio-in-a-box system (some computers also have a similar function). Why do you need to find the zero crossing? If you don't edit at a point where there is no volume, you will get an audible pop or click between the new files. Zoom in on your computer screen to see this better.

FIGURE 12-3 shows a magnification of a screen showing a zero crossing. See where the wave hits the middle line in the center of the picture? That's a zero crossing that Pro Tools found automatically. Cutting your audio files at this point ensures that your edits remain seamless and undetectable.

FIGURE 12-3

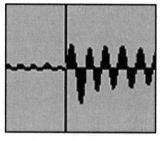

◀ Zero crossing
Screenshot used by permission of Digidesign/Avid Technology, Inc.

Punching In

Do you really think you're going to hit the record button at exactly the right point, and hit stop at the end . . . exactly in time? Survey says, no, probably not. Have no fear; punching in is here to save you. Punching in is simply automating pressing the record button. Every DAW and studio-in-a-box has a function for automating the record process of punching in. You simply tell it where to start and stop recording and it takes care of the rest for you. Look at your manuals for your system to find out specifically how to do this on your equipment. Automated punching is a key feature on digital systems. If you work alone, automating the recording process is essential to working efficiently. It helps separate the engineer from the musician.

QUESTION?

What do I do if I can't find a good point to cut my file?
You might not find a natural place to punch in. Certain performances are very hard to edit this way. When in doubt, try recording the whole section again.

Cross Fades

If you are planning on doing a lot of punch editing in your music, you will love cross fades. As you might have already noticed, even if you get the correct place to edit, chop your file up nicely, and add in the part, getting the volume levels perfect between the old and the new can be difficult. Sometimes this can cause the new part to stick out a bit. What you need is a cross fade. Take a look at an edit with cross fades in **FIGURE 12-4**.

FIGURE 12-4   ▼ Cross fades   *Screenshot used by permission of Digidesign/Avid Technology, Inc.*

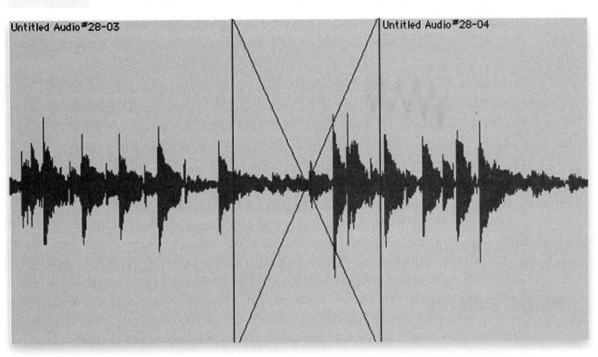

A cross fade is an automatic volume change. The very last few millisec-onds of the first file gets its volume dropped down while the beginning of the next file starts low in volume and comes up to normal. This volume exchange between the edit points makes a world of difference. It prevents any abrupt shifts in volume from being noticeable. Cross fades are an indispensable part of editing. Learn to use them effectively. You will find cross fades on comput-ers and some studio-in-a-box systems.

Combining Performances

Since we're big on examples here, let's give another one in the category of "combining performances." Let's say that you recorded a band recently. You used multiple-miking technique with great isolation—edits are possible. The band recorded three straight takes of the same song. Unfortunately each time, a different person made mistakes along the way. On the first take, the guitar solo was a piece of art. On the second take, the vocalist nailed his part, perfectly in tune (a rare occurrence). On the third take, the drums were better than on the rest of the takes. So what do you do? Go for a fourth one? Nah! Edit!

Certain things have to be in place to make this kind of edit, which, by the way, is done all the time in professional studios. First, you need to have very good isolation when you record, otherwise when you piece the parts together you might get leftover bleed in certain tracks that you can't get rid of. Second, the band needs to play to a click track, which is a steady metronome-like pulse that keeps the band from speeding up or slowing down in order to have very consistent time. If each take was performed at a different tempo, you'll have a hard time fusing them into a super-take. If all of this works, it's pretty easy to cut out sections and glue them together. This can be done easily on both studio-in-a-box systems and computer DAWs. Try it one day; it just might come in handy.

MIDI Editing

MIDI has had powerful editing from its beginning. Audio has played catch-up to MIDI for years, so it's really unfair to compare the two. Audio is a complex waveform that is hard to replicate digitally. MIDI is simple text commands. Let's get into what you can do with MIDI editing.

Quantize

Since every MIDI note is a separate event, taking control of single notes and moving them around is quite easy. One of the things you can do with MIDI is called quantizing. Quantization sets up a rhythmic grid for all of your notes to follow. If you're recording a MIDI drum part that needs to be right in time, it can be hard to get it recorded right; the quantize feature can help. Quantization pulls notes that are slightly ahead or behind right onto the beat you tell it to. It makes parts rhythmically very exact. The only information you need to quantize is the speed of your fastest note division. It uses the standard musical note durations of whole, half, quarter, eighth, sixteenth, and so on. You select the quantize value (your fastest note), and it will make a virtual grid for all of the notes to cling to. It will make the part play exactly in time. Quantizing can really steady up recorded performances.

FACT

Like to swing? Many quantize functions allow you to align notes to a swing grid. A swing grid is another type of quantization that allows for a jazz or swing feel, which is different from a straight rock groove. Swing quantize is great for jazz drum parts that need to swing!

Humanize

One of the unfortunate side effects of quantizing is a "stiff" rhythmic feel. Let's be honest, no one plays every single note right on the beat, no matter how good they are. Quantized tracks can sound too perfect sometimes.

To combat this, many sequencers have added a humanizer preset, which randomly moves selected notes off the grid, ever so slightly, to simulate the imperfect performance. It does so subtly; it doesn't sound wrong. The slight imperfections in timing that it produces can take the mechanical feel out of quantizing. Some programs call the humanize command the randomize command.

Groove Templates

A groove template is another kid of quantization. Instead of drawing your notes to a mathematically placed grid like quantization normally does, groove quantization has a preset grid that creates a nice "groove feeling." Your notes are drawn to a preset grid that is slightly out of time at certain points. Unlike humanizing, groove quantizing is preset, while humanizing is random. Again, this can add some life to otherwise mechanical sounding performances.

Looping

Do you have the same one-bar drum pattern repeating throughout most of the song? Don't play it fifty times in a row; play it once and loop it in the sequencer. This will save you a lot of time and energy. Looping is like cutting, copying, and pasting in your word processor.

Transposing

Feel like changing the key of your song after the fact? Or maybe the singer you brought in has a lower range than you thought. Your work is not lost.

MIDI is very easily transposed into different keys. On your sequencer, highlight the section you want transposed and tell the sequencer how many semi-tones (or half steps) up or down to move it. Instant transposing.

Tempo Changes

MIDI adheres to tempo maps, and every sequencer has tempo indications. It's very easy to speed up or slow down a performance you've already recorded by just changing the song's tempo. You can do this globally for the whole track, or you can create tempo changes for only certain parts. You can even change the tempo while the sequence is playing back. Since each program is a little different in how it treats tempo and tempo changes, refer to your documentation for more specific instructions. Most programs have the tempo indicator right next to the play, pause, and rewind button, which is called the "transport." The transport is where you control pause, fast forward, rewind, and record. Usually, tempo is coupled next to the standard transport controls. You can even set up different meters such as 3/4 time for the first ten bars, and 4/4 time for the rest. With MIDI, anything is possible!

Chapter 13

Tweaking Your Sound

O nce you've laid down your recorded tracks, it's time to start adding effects and mixing to achieve a final, polished sound. Proper mixing can make even the simplest recording sound great. Effects add a layer of magic dust and final smoothness to recordings. Without proper mixing and effect placement, you won't get the sound that you're used to on professional recordings. This is one of the most important parts of the recording process!

Essentials of Sound

As a recording engineer in training, you'll have to know a little bit about sound waves and electricity, because they are pivotal to understanding recording. In this chapter, you'll see why it's impossible to separate music from science—the terminology is everywhere, impossible to escape. Have no fear!

Sound Waves

Sound is emitted by a source and travels in waves that vibrate back and forth pushing air molecules around them. The sound waves create sound pressure (volume) as they push through the air molecules, which make our eardrums vibrate and pick up sound. Without a medium for sound waves to travel through, there is no sound.

FACT

Sound needs a medium to carry its waves: air, water, and the earth, itself, can all transmit sound. In space there is no sound, because in a vacuum, there is nothing to transmit sound.

The speed that a sound source (a monitor speaker, for example) vibrates tells you the frequency of the sound that comes out. If a speaker is playing a perfect A (440Hz) tuning note, such as one found on metronomes and tuners, it is vibrating back and forth 440 times a second. The faster the source vibrates, the higher the sound; the slower it vibrates, the lower the sound, or pitch, you hear. Sounds are rarely made up of just one frequency; actually there are many frequencies present in any one sound. The science behind it is beyond the scope of this book, but just understand when you play or sing one note, there's more than just one frequency present.

You might be saying to yourself, why do I have to know this? That's a legitimate question, and here's the short answer: Understanding frequency and how sound works is essential to mixing and most all effects. We don't just talk about "low sounds"; you'll see on your EQ knob that "low" might have "80Hz" next to it. Your microphone might have a "100Hz roll off" on it. You might read an article about boosting the 10kHz band to improve presence and clarity. Wouldn't you like to know what that all means? Simply put, the audio community, which you are now a full-fledged member of, deals with terms like hertz and kilohertz, so you should simply learn what they mean to avoid confusion!

Ranges of Sound

Let's talk a bit about the ranges of sound you might be used to. Your stereo might have a bass and treble knob. These knobs are used to boost or cut a certain range of frequencies. The specific range of frequencies involved will differ from system to system, but this is generally known as equalization (or EQ). EQ is simply the boosting or cutting of certain frequencies of a sound. The most basic EQ you will encounter is a three-band EQ on a mixer (either outboard or virtual).

FIGURE 13-1

 Mixer channel strip

As you can see in **FIGURE 13-1**, there are values in Hz next to the knobs. The values show what ranges of sounds are affected by turning those knobs. If you really want to learn about EQ, twist knobs and listen. Like any other skill, you need to practice. Don't be afraid to grab knobs and listen to what happens.

Effect Types

There are a few different types of effects that are utilized in studios. The first, equalization (EQ), isn't really an effect per se, but for our purposes, we'll lump it in with the rest. EQ comes in many flavors, from a simple three-band EQ found on many 4-tracks and mixers to elaborate parametric equalizers that give a great deal of control over individual frequencies. Dynamic processing involves effects that control the volume or dynamics of sounds. Effects like compression, limiting, gating, and expanders all control the volume of tracks.

Special effects usually encompass delay and its many incarnations, such as tape delay, and multitape delay. Modulation effects like chorus and phasers and flangers change the sound by utilizing a delayed signal mixed in with the original signal, which either delays that signal or changes how the delayed signal sounds. The mixture of the two signals is the characteristic sound of modulation effects. Reverb is the most important effect to learn how to utilize well. Every sound we hear has some reverb. Reverb, which is short for reverberation, is a natural occurrence when sound waves reflect and bounce off surfaces. The larger the room, the longer it takes the sound to come back to your ears—giving you the feeling of space and distance. Reverb is such an important part of acoustic sounds that when we record without it, it sounds quite strange. It is possible to emulate the sound through reverb processing, giving you the feeling of space.

Hardware vs. Software

Years ago, effects were done exclusively by outboard effects units that were rack mounted. Certain effects processors were multifunction units and could produce reverb, delay, and other effects all within one unit. Other gear was more specialized to one job, like a compressor for instance. The great part about outboard gear is that it sounds really good. Many studios still use them instead of computer plug-ins because the engineers feel the sound is better.

Early outboard gear used analog technology to produce the effects. As technology improved, manufacturers turned to digital signal processing (DSP) chips to improve the quality of the sound. The digital-effects processor was born. It was only a matter of time before a computer was able to do the job of DSP. Indeed, that day has come. Now, instead of needing floor-to-ceiling racks of gear, you can re-create all the effects you want through software. This is where the home studio became powerful. No longer do musicians need all the space and expensive gear! Now, through software, a computer (or a studio-in-a-box) can do it all.

Even with the innovations of software plug-ins, many professional engineers opt for tried-and-true hardware devices over plug-ins. Some of this is habit, and some of it is because hardware just sounds better.

Equalization (EQ)

Equalization is simply the boosting or lowering of certain frequencies, or groups of frequencies, that are present in a sound. Equalization can have a dramatic effect on a sound. However, EQ can only boost or lower what's already there; it can't add frequencies that aren't already there. In a recording, EQ helps balance the sounds between groups of instruments and alters the color of individual tracks so that they either stand out or fall back in the mix. EQ can also fix problems such as proximity effect and noise in recordings.

There are several different types of EQ that you can use to help equalize the sounds in your music. Here are the terms you might run into:

- **High pass:** Lets the high frequencies pass and blocks the low frequencies
- **Low pass:** Lets the lows pass and defeats the highs
- **High shelf:** Controls a specific frequency and all frequencies above it (similar to your stereo's bass and treble knob)
- **Low shelf:** Affects a certain frequency and all the others below it
- **Parametric:** Lets you boost or cut a specific frequency (to boost just one little part of a sound)

FIGURE 13-2 shows each kind of EQ listed, using a Waves computer plug-in.

FIGURE 13-2

▼ Types of EQ. Top row, left to right: high pass filter, low pass filter; middle row, left to right: high shelf, low shelf; bottom row: parametric EQ

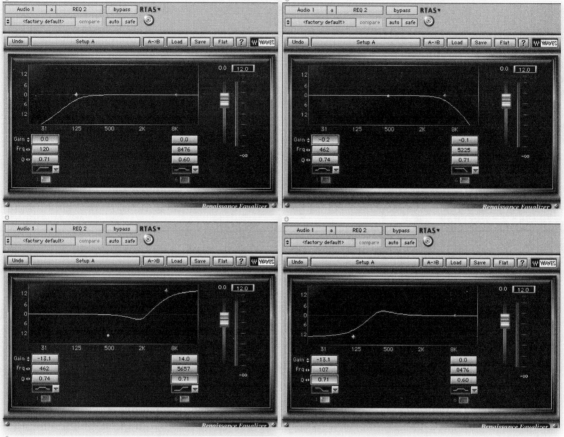

All 5 screenshots used by permission of KS Waves, Ltd.

The Parameters of EQ

Thankfully, EQs don't have a ton of parameters to set, so let's go through them one by one. **FIGURE 13-3** shows a plug-in window from Digidesign Pro Tools.

FIGURE 13-3

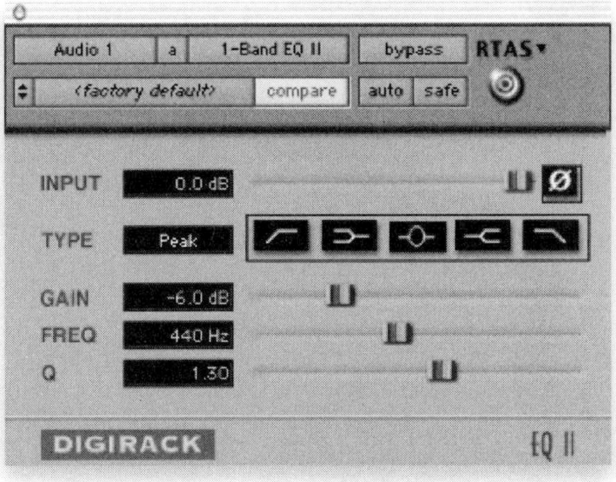

◀ Pro Tools EQ plug-in
Screenshot used by permission of Digidesign/Avid Technology, Inc.

As you can see in this figure, there are only a few things you can set:

- **Input:** This is where you can set the level of incoming signal. You wouldn't usually change this, because you don't want to overload the plug-in.
- **Type:** This is where you choose from the five types of EQ we listed earlier. From left to right, the symbols are high pass, low shelf, peak (parametric), high shelf, and low pass. These same symbols are used in most EQ software or units.
- **Gain:** This is where you select whether you are boosting or cutting frequencies. 0dB is no change, −dB cuts, and +dB boosts.
- **Frequency:** For each of the five types of EQ, this setting has a different meaning. On a low pass, the frequency you set here determines where the cutoff starts. In the parametric or peak EQ, the frequency identifies the specific frequency that will be boosted or cut.
- **Q:** The Q value widens or narrows the frequency range that's affected. Take a graphical look at both a high Q and a low Q using the Waves EQ, shown in **FIGURES 13-4** and **13-5**.

FIGURE 13-4

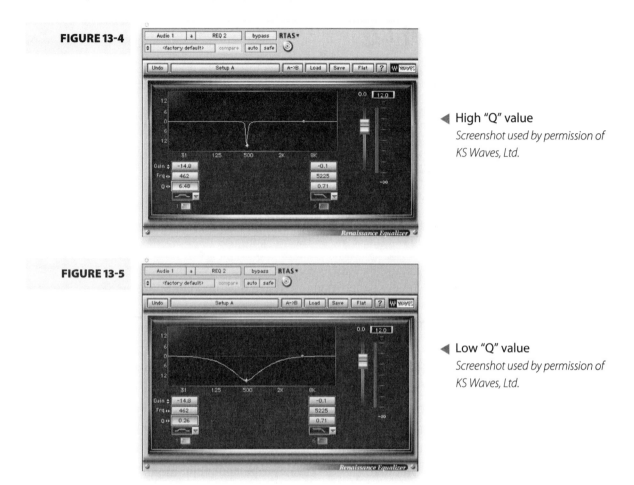

◄ High "Q" value
Screenshot used by permission of KS Waves, Ltd.

FIGURE 13-5

◄ Low "Q" value
Screenshot used by permission of KS Waves, Ltd.

Knowing When to Use EQ

There are many uses for EQ. You can use it to enhance the bass in dance/electronic music. You can use it to add sheen to vocals by boosting some high range. You can take the thump out of bass drums by lowering the low end. You can help an instrument come out the mix more by enhancing certain frequencies. You can eliminate proximity effect by using a high pass to eliminate the bass rumble associated with proximity effect. The list goes on . . .

But EQ is not a magic cure-all for sound problems. Actually, most of the time you shouldn't need to EQ much at all, because much of your EQing comes from microphone placement. It won't fix harsh-sounding instruments or poor microphones. It can certainly help things, but over-EQing will sound unnatural.

Graphic EQs show a visual depiction of sound frequencies. Red lights pop up and down along with the frequencies present in the recording. This makes a great demonstration of how different instruments sound and what frequencies give those sounds. Being able to see with your eyes will help you make a connection to your ears. Listen to your own recordings as well as to professional CDs to learn what a good, smooth mix sounds like.

Reverb

No other effect is as natural as reverb. Every sound reflects off surfaces and comes back to the listener. The time it takes to do this creates a feeling of aural space. A great concert hall is made specifically to control the reflections and provide a rich warm reverberation. Electronic devices, hardware, and software try to emulate this sound. In recording, it's one of the most commonly used effects, and almost universally necessary for some instruments, especially voice. When mixing, reverb has the effect of bringing certain sounds to the foreground or background of the mix in addition to creating a natural ambience.

Reverb is a really fast echo. However, there are many different types of reverb today. The types fall into two categories: room emulations and "old school," like plate and spring reverb. Room emulations try to re-create how the sound reverberates in different size rooms. The larger the room, the larger the natural reverb you'll get. You typically see "small room," "medium room," and "hall" as popular reverb choices. In the early days of recording, reverb was simulated by sending the audio either through a spring or on a large plate of metal to simulate the sound of reverb. "Spring" and "Plate" reverb have their own distinctive sounds and are now emulated by modern reverb processors and plug-ins.

The Parameters of Reverb

Unlike EQ, reverb has many parameters to deal with. Let's look at the D-Verb plug-in by Digidesign (see **FIGURE 13-6**) to see what kinds of control it gives you. Effects built into studio-in-a-box and hardware will have similar, if not identical, parameters.

FIGURE 13-6

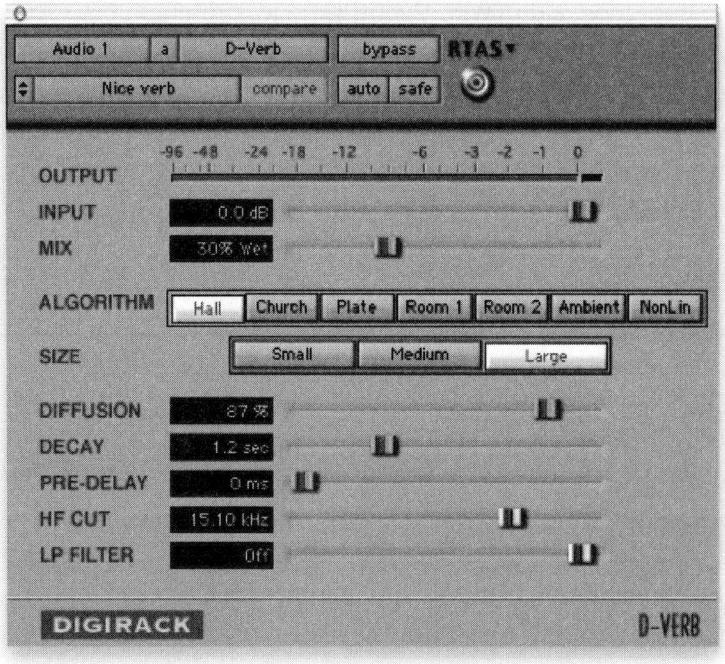

◀ DVERB reverb plug-in
Screenshot used by permission of Digidesign/Avid Technology, Inc.

- **Input:** This is where you set how much signal comes into the reverb plug-in.
- **Mix:** Specifies how much of the affected signal is mixed in with the original. One hundred percent wet and effected signal will sound very odd, as though you're in another room very far away, while lower values will incur less reverb. Play with the mix until it sounds natural and good to you.
- **Algorithm:** The algorithm is the computer's model for different rooms and means of reverbs. Each algorithm will sound different and has unique characteristics. You'll find a few you like.
- **Size:** Based on the algorithm you choose, you can set how large the emulated room is. The bigger the room, the more echo and reverb you get.

- **Diffusion:** Diffusion simulates a room's reflectivity. The higher the diffusion setting, the more "live" the room will sound.
- **Decay:** How long will the reverb hang around? If you set it long, the sound will bleed around and get very "mushy." You can get some cool effects by playing with the decay. For a natural sound, don't set it too high. A decay time under one second will do nicely for most things.
- **Predelay:** Since reverb takes a short time to appear in a natural environment, the predelay is useful in making your sound feel real and natural. Experiment with the predelay time to find out what suits you best. For most situations, fifteen milliseconds to forty milliseconds will sound the most natural.
- **HF cut and LP filter:** High frequency cut and low pass filter are holdovers from our EQ discussion. In a reverberated room, high and low frequencies react and decay at different rates. Every room will yield different results. Usually adding some of these parameters will help your reverb sound more natural.

Every reverb processor includes presets—explore these to get an idea of what you like and how they work. Pay attention to how the different algorithms, room sizes, and decay rates influence the overall sound. Tweak to your heart's content—you might be able to improve the presets and make them your own.

Knowing When to Use Reverb

Reverb, in some subtle way, is used on almost every part of a recording. This is not to say that you're going to apply gobs of reverb everywhere. Reverb is one of those things that when applied well is hard to detect. Emulating a natural sound shouldn't draw attention to itself. If you can hear the reverb, unless you're going for a special effect, it is a bad thing. It's a mistake many make because reverb tends to make things sound really good. It's a bit of a drug and it's hard to stop, but go easy.

Certain instruments need reverb more than others. Bass typically has little reverb applied to it. Guitar will need some reverb—especially acoustic guitar. Vocals will almost always get reverb; vocals without reverb will sound dry and unnatural. Drum sets tend not to get reverb applied to the whole set.

However, snare drums typically benefit from some subtle reverb. Because of how cymbals ring, it's not wise to put reverb on; the same goes for bass drums. Acoustic pianos should not need much reverb, but a subtle amount might warm things up. As for other instruments and other situations, trust your ears.

ALERT!

Reverb is a great effect, but it's easy to get carried away. Here is a very easy way to set reverb well. Start by applying a reverb on an auxiliary channel. Start mixing in the reverb until you can hear that it's there. Then, back down slightly. It seems easy, but if you can hear an effect, you are probably overusing it. Dial it in, and then back off a touch.

While you could make an entire album with just EQ and reverb, there are other essential effects that engineers use to craft recorded tones. The next chapter will detail everything else you need to know about effects, including how to get them into your recordings.

Chapter 14

Using Other Effects

Beyond the basics of EQ and reverb lives a whole other world of effects. Chorus, delay, distortion, and various other effects help shape your sound in unique ways. The little things will put your recording over the top! This chapter also covers what you need to know about the techniques engineers use to apply effects to music.

Chorus

Chorus is a commonly used effect. The term "chorus" comes from the idea that, if someone were playing the exact same part as you, he or she wouldn't be exactly in tune and exactly in time with you. The delay might be only several milliseconds, but that's enough to create an effect of multiple players. It makes it sound as if more than one is playing; the end result is that they sound richer. Chorus replicates this by copying the signal, delaying it a bit, and detuning it through a modulation effect. Modulations are changes to the pitch that rise and fall in a steady pattern. The change in pitch gives chorus its distinctive sound.

FACT

Chorus produces its modulation through a low frequency oscillator (or LFO). The resulting sound is a shimmery effect that replicates doubling of an audio signal.

Parameters

The amount of control you get from chorus varies depending on your equipment. A traditional chorus effect will give you the control of these elements:

- **Delay:** Controls how long it takes for the second copied signal to appear. The amount of time is generally kept fairly low—usually between fifteen and thirty milliseconds.
- **Depth:** Controls the amount of change in modulation or pitch of the sound. The higher the number, the weirder it's going to sound!
- **Rate:** Controls how fast the pitch will rise and fall.

Some choruses really go to town. Take Waves, for instance; they make MondoMod, the "Rolls Royce" of plug-in effects. Take a look at how much control MondoMod gives you (see **FIGURE 14-1**). The extra control can yield some incredible sounds.

FIGURE 14-1   ▼ Waves MondoMod   *Screenshot used by permission of KS Waves, Ltd.*

The parameters for chorus must be tinkered with. It's impossible to give you standard chorus presets because everyone will use them differently. Play around and have a good time. Check the included presets, too.

Uses for Chorus

Chorus is used on many different sounds. Guitar players love it for creating clean sounds. Keyboard and synth players commonly use it to thicken up their sound. For vocals, chorus can help cover up subtle pitch problems. Chorus can also add the illusion of width to a sound, making it appear fuller and wider. As always wherever you use chorus, go easy and don't go overboard.

Compressing

Dynamic effects change and control the aspects of volume in a recording. While not as dramatic as reverb and chorus, dynamic effects are some of the most important tools in recording—and the ones that are easily forgotten or misused in home recording. Dynamic effects come in a few flavors: compressing, limiting, and gating.

ALERT!

Dynamic effects are subtle and don't alter the character and tone of a sound; they affect only volume, and only in a subtle way. If you slap on a reverb, you can instantly tell what's going on. Something changes right away. Grab an EQ knob and you impart some change immediately. But with dynamic effects, you get a less obvious change.

A compressor is an effect that automatically stops volume from rising too high. When instruments spike their volume wildly, as drums and bass do, for example, the sound can be unpolished and hard to listen to. A compressor, when set correctly, keeps sudden volume changes from occurring. It compresses the loud signals, makes them less noticeable, and smoothes out the overall volume level of a track.

Compression has many uses in the studio. It controls instrument levels when you're tracking. It helps bring tracks to a smoother volume level when you're mixing. It also compresses the entire song to smooth out the volume when you're mastering.

Let's take a drum track, for instance. Drums produce a wide range of sounds, from very soft to quite harsh. In other words, their dynamic range is pretty wide. In order to set the volume level correctly, you need to account for the loudest hits and make sure they don't clip the channel. This can be a challenge and usually means turning down the track, which can make the softer parts harder to hear. Applying a compressor limits the loudest hits from getting too loud and lets you raise the overall level of the track without fear of clipping. This helps make the drums fit into the mix better and also serves to smooth out the sound. This is one of the most common uses of compression.

Parameters

A typical compressor has a few controls that set how the compressor acts. **FIGURE 14-2** shows the Waves Renaissance Compressor plug-in for a visual example.

FIGURE 14-2

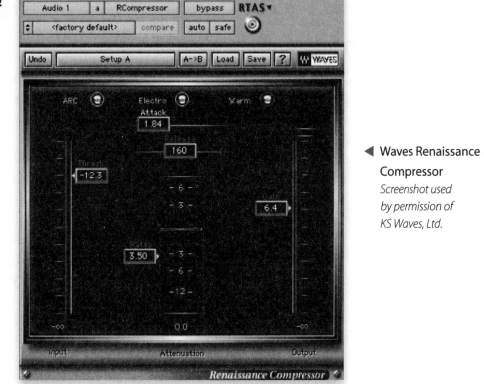

◀ Waves Renaissance Compressor
Screenshot used by permission of KS Waves, Ltd.

- **Gain:** Sets the overall volume of the plug-in. When using a compressor, it's typical to boost the gain because you will have more volume range at the top to work with.
- **Threshold:** Sets how loud the incoming signal needs to be before the compressor starts to work. The lower the threshold (in negative numbers), the more signals get squashed.
- **Attack:** Controls the length of time (measured in milliseconds) the compressor takes to actually start to change the sound level after the volume reaches the threshold.

- **Release:** Controls the length of time (in milliseconds) the sound should be held by the compressor after the volume level falls below the threshold.
- **Ratio:** Here is the most important parameter! The ratio is the difference between incoming and outgoing signal. A 3:1 ratio says that when 3dB of sound come past the threshold, an increase of only 1dB comes out the other side. Ratios of between 2:1 and 4:1 will sound natural. Any higher and the dynamic range may sound "squashed" and quite unnatural.

How to Use a Compressor

Setting up a compressor isn't hard; you just need to be aware of a few parameters that help it to sound natural. The threshold should not be set so low that the compressor is always working. You really want to compress only the upper end of the dynamic range. The ratio depends on how much fluctuation the source has. If it's a snare drum, you might want a 4:1 ratio or even higher. If you're trying to lightly compress a guitar track, choose a lower ratio. Attack and release are where compression is won and lost. Attack is not as critical as release. If the release is long, then the instrument will sound unnatural. You have to listen for the instrument's natural volume decay and try to match that with release time. As mentioned before, compression is not a dramatic effect, and it might take you quite a while to hear if it's even there.

Try bypassing your effects often to hear the before and after material. This way you can judge how you're doing. Bypassing effects is commonly referred to as A/B testing. ("A" is with the effect on, and "B" is without any effect.)

Limiting

A limiter ensures that no signal gets too loud. It sets a "do not pass this" level, and anything that gets near that level gets squashed. A limiter is actually a *super compressor* with a ratio of 10:1 or more, and many limiters use a ratio as high as 100:1! Limiters are good for turning up an entire track via the gain parameter with absolutely no fear of clipping.

The controls of a limiter mimic the controls of a compressor because that's what limiters are—compressors. However the ratio of a limiter is higher than that of a normal compressor, or it may be preset. Because of the dramatic squashing effect that limiters incur, you should use them carefully and only on sources that really need it. Compressing is usually used more often than limiting. Limiting is used often in mastering to get a full, loud signal.

Gating

A gate, or a noise gate, is basically a backward compressor. While a compressor limits the high end of the dynamic range, a gate quiets the low end. For example, when an instrument is not playing, there is a break in the sound. If the drummer doesn't play for the first chorus of the song, the microphones might be picking up other instruments or noise. Instead of tuning the volume down manually and raising it up again, a gate can be set to close the channel until a certain volume level is achieved. This can cut down on noise and bleed from other sources. **FIGURE 14-3** shows a gate plug-in from Pro Tools.

FIGURE 14-3

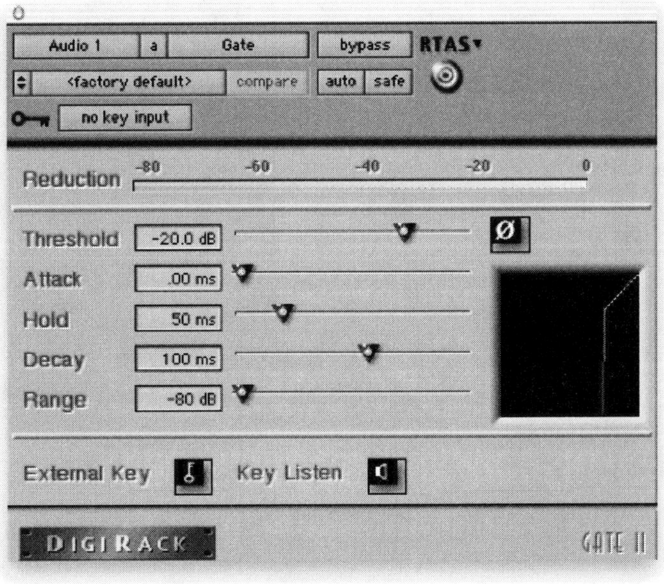

◀ **Noise gate**
Screenshot used by permission of Digidesign/Avid Technology, Inc.

You'll notice that the parameters are very similar to those of a compressor.

- **Threshold:** Sets the volume level at which the gate opens and closes. The gate will open when a real signal is present at that level and close when the signal falls below that level.
- **Attack:** Sets how fast the gate acts after the volume reaches either the open or close threshold.
- **Hold:** Sets how long the gate stays open after it falls below the set threshold.
- **Decay:** After the hold releases the signal, how long until the gate fully closes.
- **Range:** Determines what -dB value "closed" is.

Experiment with gate settings to get rid of noise and other low-level volume problems in your tracks. You can even use the gate for some creative effects—your creativity and tinkering are unlimited!

FIGURE 14-4   ▼ Wave Arts' TrackPlug   *Screenshot used by permission of Wave Arts, Inc.*

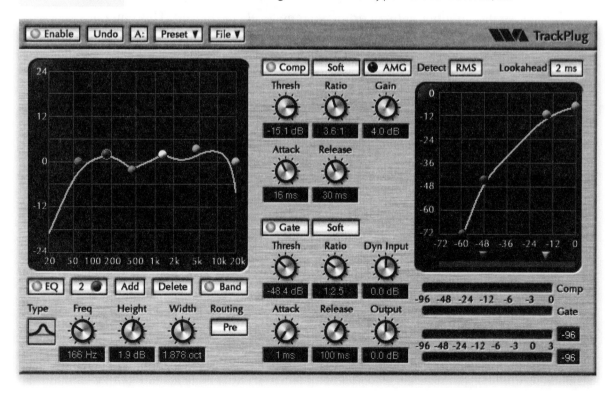

On a large mixing console, it's not uncommon to find EQ, a noise gate, and a compressor on each channel. Because these are the most commonly used effects, it's useful to have them hard-wired. Plug-ins also combine EQ, compression, and gating into one plug-in. These plug-ins are so useful, you might even use them on every track in your mix. **FIGURE 14-4** shows a shot of TrackPlug by Wave Arts.

Other Effects

There are even more effects you can use! The basic, most common effects are covered here, but feel free to experiment with other effects now that you have a good idea what the parameters do.

Distortion

Remember how we said that overloading a circuit is bad? Well, most of the time it is, but ask any guitar player and he or she will tell you that distortion is good. Distortion amplifies the incoming signal wildly, letting it distort, but this doesn't incur "bad" clipping like overloading a recording channel does. When you overload a recording channel, you ruin the recording; distortion, instead, gives the effect of overloading a circuit on a guitar amp. It can make for an interesting effect when used! It's a great way to add "dirt" to a sound.

Delay

Delay is a simple effect that copies your signal and re-creates it, only it does so at a specified interval after you play. This can range from short delays in the milliseconds to long delays of many seconds. A short delay will thicken up a track much as a chorus will. A long delay can have many musical possibilities. If you set the delay right, the delay will follow you around, playing back what you played. Delay has long been a favorite of guitar and keyboard players. Try it on your tracks to see how it sounds.

Noise Reduction

Noise reduction is one of those amazing effects that once you have it, you can't live without. Unlike a gate that limits noise when you're not playing, a noise reducer listens to the signal and gets rid of all unwanted noise. Hums and hisses? Gone. Crackles and pops? Gone. Simply talking about Waves X-Noise plug-in can't possibly convey how amazing it is. It knows the sonic fingerprint of different noises that the designer has programmed in.

The most incredible button to press is the difference button. That button isolates only the noise it took away so you can hear what's no longer there. It's amazing to hear all the high frequency hisses and buzzes just magically vanish. For those of you who rely on hardware processors, check out the Hush line of noise reducers—they are equally impressive.

Adding Effects to Your Recording

Now that you know what the effects do, let's actually get to the nitty-gritty of how to use them. Effects can be added in various ways to your recorder. The common ways to add effects are through inserts, auxiliary channels, and patch-throughs.

Insert Effects

Back in the discussion on cables in Chapter 8, you learned about an insert cable, called a tip ring sleeve (TRS), with a stereo cable on one end and two individual mono plugs on the other side. This is the standard insert cable. An insert effect is any effect that you add to one specific channel only. Your recorder or mixer will most likely come with some inserts. Basically what happens is that the stereo cable acts as a send-and-receive device for the effects unit. The cable has two wires—one sends your signal to the processor, and the other receives the effects signal and mixes it back into your sound.

An insert sends your entire signal to the processor and mixes all of the signal back. Insert effects work well with dynamic effects (compressing, limiting, and gating), which take your entire signal and affect the total volume of that signal. Since there is no need with dynamic effects for mixing how much effect you use, inserts work perfectly with these effects. But because you have no control over mixing when you use inserts, they aren't appropriate for all the effects. Effects like reverb and chorus, for example, would be overpowering because you couldn't control how much of the reverb or chorus is mixed with your signal.

FACT

In software programs, insert effects are usually listed above the virtual mixer's fader for each individual channel. This varies from program to program, so check your manual to be sure.

On a mixing board, EQ is a good example of a "hard-wired" insert effect. All of your signal passes through the EQ and then gets mixed out to the stereo pair.

Auxiliary Channel Effects

One downside to insert effects is that if you use a piece of hardware on one channel, you can't use it anywhere else. So, for example, if you need compression on four channels, you need four hardware compressors! On the other hand, if you use plug-ins on your computer—no big deal—you can use as many copies as your computer will allow.

Auxiliary effects plug into the appropriately named auxiliary inputs on your mixer, recording device, or software auxiliary track. They use similar send and receive cables that insert effects do; the only difference is that each channel (or track) has its own blend knob for setting how much of the effect to mix in with the original signal. This makes auxiliary channels perfect for reverb and other processors that you want to assign to more than one track. You can also get away with having one hardware effects processor to cover many of your needs. Most mixers have at least two aux sends, allowing you to patch in two separate processors. The bigger mixers let you do more. On a computer you can "aux" to your heart's delight, or until your CPU gets overwhelmed. The studio-in-a-box systems supply auxiliary channels in tandem with hardware and software. You can use their effects and virtually mix them together inside the unit, or you can place hardware processors in the auxiliary inputs and use them. This is a very flexible way to go.

If your computer interface has enough inputs and outputs, you can do this with a digital audio workstation (DAW) as well.

Patch-Through Effects

Usually you want to record instruments without any effects on them. This allows you much greater control later on when you mix. If you record with effects, the effects are "printed" to the track; you can't get rid of them. But there are some circumstances when this is okay. Guitar players, for instance, tend to use reverbs, delays, and choruses to make up their signature tone. It would be hard to replicate their exact sound later, so it might be best to allow them to record with their normal effected sounds. Keyboard players also do this on their synthesizers—adding effects into their patches. In these cases, it's okay to let the players "patch-through" their effects.

The Pros and Cons of Software Plug-Ins

It's no secret by now that the computer and plug-ins are becoming the cornerstones of the home-recording market. Computer effect plug-ins have specific advantages over hardware. Here are some of the pros:

- You can use them as many times as your computer's processor can handle. For example, you can have five different versions of the same reverb, on five different tracks, or aux tracks—all with different settings on each plug-in.
- You gain the ability to use one plug-in as an insert effect or an auxiliary effect at the same time.
- You can conceivably have a compressor on each track, all for the price of one.
- You can save and recall patches from song to song automatically.
- You won't need cables, patch bays, and rack mounts—there's no space required, except virtual space.
- You might get to upgrade your plug-in version for free if the maker updates it to improve the sound.
- You can automate the settings of any plug-in within a song.

There's always a downside:

- Just because you bought a plug-in doesn't mean you'll be able to run it.
- The more plug-ins you use, the slower your CPU becomes, and the fewer tracks you can use.
- Many professional engineers believe that the sound isn't as good as hardware versions.
- You can't use plug-ins anywhere else but in a computer, which makes hardware versions more handy for live sound or instruments.

As you can see, there are ups and downs to plug-ins, but most people will agree that the flexibility and cost-benefit ratio make computer systems and plug-ins very attractive.

Chapter 15

Mixing and Mastering

Mixing and mastering are two of the hardest subjects to describe in writing. By its very nature, audio is an art of the ears—not the eyes. If you spend enough time in studios, talk to enough engineers, and read through the important work on the subject, you'll see some similarities in how engineers work and how they mix. But there will never be a substitute for getting your hands dirty and mixing yourself.

What Is Mixing?

Mixing is the second stage in the recording process, which comes after the tracking has completed. In a basic sense, mixing isn't that hard to understand. Mixing involves blending all of your separate tracks into one stereo pair suitable for listening to on any radio, Walkman, iPod, or car stereo. Mixing also involves adding effects to polish up the sound.

The art of blending disparate sounds is very difficult. When you hear an acoustic band, the blend is taken care of for you; the reverberation is natural from the room. As soon as you start close-miking instruments, reproducing the sound in a realistic fashion becomes a challenge. While you might not know how to make a good mix yet, you certainly know a bad one when you hear it.

Mixing is all about perception. Can you perceive that this group of instruments really sounded this way? The best mixes sound natural, and they try to replicate how those instruments should blend together. If the mixing engineer has done his or her job, nothing out of the ordinary should be noticeable. That is, nothing catches your ear as "unnatural" or out of place. As you know, it's easy to spot a bad mix; there's just something "not right."

Many engineers talk about hearing in multiple dimensions. Understanding those dimensions can help you figure out what's going on in a good mix. Here are the basic dimensions you'll encounter in mixing and what it means to work with them:

- **Foreground/background:** Bringing sound forward and backward in a track using volume
- **Depth:** Using effects to create the feeling of closeness or distance
- **Up and down:** Using EQ to help tracks sit in their own distinct part of the frequency spectrum
- **Side to side:** Placing sounds from left to right using the pan controls

Without oversimplifying the process too much, these four dimensions give you an idea of what goes into a mix. Now let's look at what goes into working with these dimensions so that you can start mixing like a pro.

Foreground/Background

The basic element of mixing is the loudness of each track. This is the first place you should start as a budding engineer. No matter what system you own, from the 4-track to Pro Tools, volume manipulation is available to you. In visual artworks, such as paintings and photographs, there are background and foreground; the more important visual elements usually come to the front of the work. It's the same with audio—the important parts need to be heard.

The mute and solo buttons are the two most important buttons to use in mixing. Get to know and love those buttons. When you mix, you'll never be able to work on a full mix all at once—it's just too much sound at once. Isolating sounds or groups of sounds played together is the best way to go, especially when working with EQ problems. Focus on small parts and build your mix around them.

The volume control on a mixing board or your software recording device is called a "fader" because it allows you to fade the sound in or out at will. The first thing you should do is set your faders for each track to create a basic feeling of foreground and background. In most music that includes vocals, the vocal track is usually the point of interest and should be the loudest thing you hear. But how loud? How much louder than the accompanying guitar? You have to trust your ears on this. At a basic stage like this, do your best to get it to sound as balanced as possible. Volume of the tracks is only one very basic element of a mix. But it's a great place to start!

Depth

The element of depth is taken care of by effects. Depth shouldn't be confused with volume. Depth is the feeling of how far a sound is from your ears and has little to do with volume. Just imagine that you're listening to a symphony orchestra. No matter how loud or soft the violins are, they will always sound closer to you than the brass section that sits in the back of the orchestra. The way to demonstrate depth is to use reverb, which we learned before re-creates the natural reverberation and ambience of a space. Something can be quite loud in the mix, but be drenched in reverb, which can make it appear far away. And using less reverb can make the sound sit right up front, as if the player were right in front of you. Other popular depth effects are chorus and delay, which are all very closely related to reverb. All of these effects help to widen the perceived music. Remember that mixing is perception, and using effects correctly can let you manipulate your mix in some very neat ways.

FACT

If you have a track that seems to get lost no matter how you mix, try using a chorus effect to thicken up the track. Chorus adds a second copy of the original signal and slightly delays it to give the illusion of more than one person playing. It's a great way to thicken up a track and help it stand out a bit more.

Up and Down: EQ

So you've spent hours manipulating the volume and depth of your track, and yet, everything sounds bad. No matter what you do, it just seems sonically cluttered, so to speak. EQ is the answer, but in a different way than you might think. Before now, we have seen EQ as a way to shape the sound of individual tracks. When you mix, EQ takes a slightly different role.

Remember the idea that any single sound is made up of many frequencies? Think of sounds as analogous to building blocks; each sound is a differently shaped building block. Not all sounds will just "fit together" without some light sanding. For example, the bass guitar and the bass drum sit in the same frequency range—the low frequencies. Depending on how the instruments were recorded, when you play them back together, you'll most likely hear a bit of sonic mud because of so much sound coming in from the lower frequencies. What's happening is that both the bass drum's and the bass guitar's low frequencies are covering each other up, making your mix very bass heavy. You'll notice that it's also hard to hear both the instruments clearly. Welcome to EQ carving!

When you mix, sounds should not compete. If you load up a mix with a bunch of instruments in the same frequencies, you'll get mud. Start to separate the sounds so that each sound can occupy its own layer. Try cutting the bottom of the bass guitar so the drum has some room. Or try the reverse and cut the bottom of the bass drum. By doing this, you open up a space for the other instrument to sit in. You make a mix much the same way you make a building—one layer on top of another. EQ helps the pieces fit together.

You probably consider a bass drum's sound as fairly low in frequency. However, whether you realize it or not, the bass's sound also consists of high frequencies, albeit very quiet ones. You can barely hear them, but they're there! To start clarifying your mix, take out the frequencies that aren't being used. If you have a lowest-frequency sound, cutting high frequencies won't have much effect on the low-frequency sound. The same is true with high-frequency sounds—you won't affect them if you cut the low frequencies. If you properly trim sound frequencies, the tracks have a better chance of sitting one on top of another without clashing too much.

Trimming with EQ is also called carving EQ. Just think about each sound and how it sits in relation to the other tracks. If you have sounds that compete, try cutting from one sound so that the other has room. Experiment, and tweak, tweak, tweak!

Side to Side

Panning is the control of sound placement. In stereo recording, music is panned between the left and right speakers. Until music is heard in surround or 5.1-channel surround sound (which new home theater systems have that utilize five speakers and a subwoofer), left and right is all we have. Just as EQ carving sets how frequencies sit on top of each other, panning controls how the sounds sit from right to left in the stereo field.

For a basic test, don't pan anything; set the pan controls exactly at 0, which means equal distribution to both the left and right speakers; this is also called center pan. Now that you've set all the sounds to the center, you have created a narrow and crowded mix. The sound is most likely muddy and indistinct. It's easy to understand that no matter how well you EQ, if all the sounds sit in the same pan position, you're going to get even more fighting over sound! Now, start moving instruments around. Put the guitar to the right, the bass to the left. That simple move opens up the mix a lot!

If you're having a hard time getting the bass and bass drum to sound good together, make sure to pan them away from each other. No matter how well you EQ the sounds, if they're sitting in the same pan spot, they will fight. Moving them around can alleviate many EQ and balance issues.

To really pan well, try to imagine a stage in your head. Think of where the sound comes from: guitar on one side of the stage, vocals in the middle, bass on the other side. Drums are often in the center but reach right and left because they physically take up a lot of space on stage. Now try to pan your mix that way. Push the guitar and bass to opposite sides of the mix and keep the vocals in the center. Let the drum mix mirror how the drums are set up— snare and kick basically in the center, a tom to the right, a tom to the left, hi-hat cymbal on the right, and ride cymbal on the left. The amount of left and right pan that you use is up to you and your ears. Each song might have very different pan ideas that suit the song, and no two mixes will be the same.

Audio in Motion

Here's something vital: Music breathes, and so should your mix. This doesn't mean just turning up the volume on an instrument for a solo! Your audio should always be in motion; even the slightest movement of volume levels makes the mix feel alive. Minute changes during the playback of the songs will do wonders, making your music feel more alive. It's a trick that so many engineers use. Keep those faders in motion, even just slightly. It will make everything sound better.

Automation

If you own a studio-in-a-box or computer recording system that supports mixing automation, you'll enjoy this section. Mixes aren't static. You can't just "set it and forget it" as the TV infomercials often say. A song and its resulting mix is a living, breathing thing that changes. Guitars get louder for solos, drums duck under vocals during verses; things change. If you're working on a system that you can't automate, such as a tape-based studio, or if you're using a mixing console, you'll have to know exactly what's going to change throughout the songs and adjust the levels and other parameters live upon playback. In the old days, this was called "playing the console" and was an art all to itself.

Digital technology allows you to plan all the movements of faders, effects, and pan settings in a process called automation. You go through the tracks as you mix, and you record the motion of the faders into a special track called an automation track. Once you've recorded all the movements, the automation track plays back and takes care of all of the changes in faders and so on. It's a great thing to have if you work alone. Most studios-in-a-box support some level of automation, and computer systems give you an incredible amount of control over not just volume, but all parameters of any effect on the system.

Fades

How will a song end? Take a good listen to the end of some of your favorite albums, and you'll realize that none of them just stop; they all have some kind of smooth fade. How do you achieve a smooth fade? If you use a digital system, you have a fade-out option in the editing menu and you can easily fade out the audio using that. If you have automation capabilities, just automate the master fader at the end of the tune. For those of you who have to move the fader by yourself, at the final mix down, just evenly yank the master fader to zero to achieve a smooth fade.

FACT

"Flying fader" is the name given to a mixing board with motors that move the faders automatically. After you've recorded automation for your songs, the fader "flies" by itself along with the volume changes you've recorded via automation. Flying faders were once found only on the most expensive mixing consoles, but now more and more manufacturers are putting motorized faders in home studio products.

Using Buses

We've talked about using bus and auxiliary inputs for certain effects like reverb, but you can use buses for other things as well. A bus is simply a path that sound can take. A good example of buses is what you can do with drum tracks. Let's say you have four drum tracks that you mixed together. The balance between the drums is perfect, but they need to be a little louder in the mix. You could try to raise all four faders equally, although you might find that difficult to do. Or better yet, you could send each of the drum tracks the same bus, which would act as a volume control for all the signals it's fed. You can then easily adjust the drum tracks together either up or down.

The End Result

Okay. You've slaved over your work. You've mixed and remixed everything and now it's time to finish up. There are a few things to do before you're done. The first is format—what are you mixing down to? If you use a tape multitrack, you'll use a separate tape player to mix down to. For the digital folks, most will mix down directly to a burned CD. Otherwise, you can mix down to a high-quality DAT tape that's suitable for CD duplication.

QUESTION?

Where do you start if you have sixteen or more tracks to mix?
While every engineer works differently, more often than not, the drums are mixed first. Each drum is singled out, starting with the kick drum. Once the drum mix is intact, add the bass. Then, you can add the other instruments. Vocals are usually last. Compression is added on a track-per-track basis on the instruments that need it. Effects are added to auxiliary channels to get a basic "rough" mix. Once the engineer has a rough mix, he or she starts tweaking EQ, setting all the effects levels perfectly and starting to craft the mix.

Final Mix Down

The final mix down entails sending all of your work to a simple stereo pair of left and right signals. All the levels and effects need to be taken care of in advance. Whatever you commit to the final file is it! If you're working with automation, get that in order too. At this point, you should feel comfortable that you have a good product. Remember the fade-out at the end if you want that effect.

Bounce to Disk

On digital systems, most notably computer systems, the process of a final mix down is achieved by "bouncing" to the disk. All the audio is bounced together onto the hard drive into a file (usually a stereo file) that can be burned. Bounce to disk is the same as mixing down to a tape or any other format. On many computer systems, bounce to disk is not something you control. This means that you can't play with levels and ride faders while it's bouncing. Many programs just do it quickly while you wait. The reason they do this is that all of the software available lets you automate movements of faders and other parameters. If you can automate, then you should. Once you have the final file, you can burn it to a CD using your favorite CD-burning software.

Dither?

CD audio is 16bit/44.1kHz audio. That's the standard of how the digital information encodes onto a CD. Every CD available today plays back that way. The newer computer audio systems advertise better-than-CD quality, and if you're getting into computer recording you might be able to record at that quality. Many computer audio systems record at 24bit/96kHz.

Without getting too technical, the higher the numbers, the better quality you get. The problem is that if you record at 24bit/96kHz, you can't just burn that to a CD, because it's not compatible with current CD quality. You have to do something called "down sampling," which means mixing down to 16bit/44kHz. In your software program, this is easy to do. In the bounce-to-disk window, you'll always be asked what rate to mix down to. If you are making a CD, you always need to mix to 16bit/44.1kHz.

Okay, so now to dither. When you "down sample," you lose some of the audio quality. It's just a fact having to do with how the computer goes from a higher sample rate to a lower one. Dithering was invented to make up for the loss in quality. If you're recording at better-than-CD quality, you need to dither your music down. On many systems, dithering is a plug-in you put on the master fader. On some systems, it's a box you can check in the bounce-to-disk window. Either way, if you record at 24bit/96kHz, or anything above 16bit/44.1kHz, you must dither it. If you don't, you can hear the difference. Your recording will sound better with dithering.

Compare

Now that you have your proud mix in your hands, you need to start playing it on as many systems as possible. Play it in your car, on your stereo, on your mom's stereo . . . you get the picture. Play it everywhere you can. The goal here is to make sure that it sounds the way you want it to sound on every system. Certain mixes sound great at home and quite bad in the car. Most professional recordings sound good on almost every system they're played on. If you find some problems, you can always go back and remix. It's also good to play the song for as many different people as you can, especially musicians. Opinions at this stage in your development are very worthwhile.

QUESTION?

How long does it take to mix a song?
Well, how long do you have? Mixing takes a very long time; you will mostly listen to the entire song at least twenty to fifty times before you get close to finishing. Now you know why bands take so long to release albums.

Mastering

Mastering is a term that's thrown around a lot but rarely understood. It's also the hardest to pull off at home. Mastering is the last stage in the recording process. Mastering takes all the separate songs for an album and puts them together so they sound good together. If mixed well, each song will sound good by itself, but that doesn't necessarily mean all the songs you put back to back on a CD will work together. Subtle differences in loudness and EQ from track to track can really hurt the sonic impact of a record. Mastering balances the sound from song to song so that the album sounds cohesive.

Mastering balances the levels from song to song, making sure that each song is as loud as it needs to be. It also makes sure that the tracks are relatively equal in volume and tone. Mastering might also apply some compression to smooth out the dynamics of the entire song. In many cases, some subtle EQ is applied to help polish up the tracks. Mastering deals only with the final stereo mix down, and it's not considered mixing anymore.

These days, mastering is done almost exclusively on the computer. There are many programs such as Wavelab and T-Racks mastering suite that let you work with just the stereo mix. You can also bring the final stereo file back into your audio program, such as Cubase or Pro Tools, to start mastering.

Mastering Tools

Compression and limiting are the most common processes applied during mastering. In order for a track to be heard well, it has to be loud enough. Even if you did a great job of setting levels in your recording, mastering sets the final loudness. Some compression might be applied to reduce the dynamic range of the audio. The negative effect of this compression is that you lose some of the volume of the track because the compression squeezes the sound together. Limiting then takes the compressed signal and boosts it to make tracks as loud as possible without clipping or going over the threshold you set. Once the track is compressed and boosted, it will start to "sit" better on the album. When this is repeated from track to track, you start getting something that sounds more like an album, rather than eight tracks thrown together on a CD.

Tone and Sequencing

When you put several songs together on an album, especially ones that were recorded over a long period of time, you might notice that the songs sound quite different from each other in regards to EQ and overall loudness. When you master, you can apply EQ to help the tonal balance of the songs fit together well. This typically means going though each song and listening for its EQ or looking at a graphic representation of the frequencies in your mastering program. Once all the tracks sound cohesive, the next step is deciding the sequence they should appear on the record. Unless you're recording a suite of songs that has a predetermined order, mastering is the stage when the order of songs occurs—this is called sequencing. Sometimes you can't help that certain songs sound different from one another, so in those cases careful sequencing can help them fit together on the album.

It's entirely possible for you to master at home, and you might even get good results. However, mastering might be the one part of recording you won't want to do at home. Mastering and a good mastering engineer are worth their weight in gold. Mastering really is an art—not to mention that mastering studios have hundreds of thousands of dollars of mastering equipment that can make your CD sound incredible.

If you've worked very hard at home, and you want to get the CD you've created put on the market, get it mastered by a professional. You'll be amazed at what a professional can do to a final mix.

Chapter 16

Getting to Work with Cubase

To really bring things together, let's take a tour of Cubase, a recording studio program by Steinberg Media Technologies. Because Cubase runs identically on both PCs and Macs, it's a solid choice to illustrate the kinds of recording software you might encounter. In this chapter we cover the most important elements discussed in this book: setting up audio tracks, adding MIDI tracks, quantizing, MIDI editing, working with loops and samples, and basic mixing and applying effects.

Why Cubase?

Good question! First, let's agree in general that demonstrating recording on a studio-in-a-box is difficult, because many of the functions take place on a small LCD screen. Demonstrating through diagrams simply won't do it, either. That leaves us with computer software, which has the best mix of functionality and graphics, as a useful example. Plus, whether you like it or not, home recording is headed to the computer.

As for why Cubase specifically? Because it works the same on Macs and PCs, and can work with any audio input hardware you choose. This is something that the other cross-platform contender, Pro Tools, doesn't allow. Also, it's a very popular and powerful software program capable of making incredible music.

This is not an endorsement of Cubase or a manual for using it. We're simply showing you all the tools we've discussed so far, and making some music with them. We're going to set up a basic mix with MIDI and audio, mix those tracks, add EQ and effects, perform some MIDI editing, and best of all, show you how it's done.

Getting Started with Your Session

Upon opening the application, Cubase gives us by default a new empty song with eight MIDI tracks and eight audio tracks. If we wanted more, we could easily add more, but this is plenty for our example. **FIGURE 16-1** shows the Cubase default launch screen. You see your tracks stacked up on the left side, the transport at the bottom, and the "playing field" in the center. The playing field is where all the MIDI and audio is going to exist for us to work with.

ESSENTIAL

If your song needs a good drum part and you don't know a great drummer, a virtual instrument just may do the trick. Steinberg's Grove Agent is a good example of this. If you are inclined to make your own drum parts, Battery by Native Instruments is a powerful and easy-to-use drum sampler/mapper. If you record all by yourself, virtual instruments can extend the dimension of your personal music.

Adding a Drum Loop

Like most home musicians who work alone, we will use some loops in our music. We've downloaded a nice drum loop from the Internet. This particular loop plays at 151 beats per minute in 4/4 time. To make the most of the session, we need to tell Cubase that we will work with the metronome setting of 151. This is easily done in the transport by clicking in the tempo box and changing the value to 151. Notice now in **FIGURE 16-2** that the transport (the remote control section) now reads "151" for the tempo.

FIGURE 16-2

◀ Setting tempo via the transport
Screenshot used by permission of Steinberg Media Technologies.

Setting up the tempo ahead of time helps us in a few ways. First, Cubase sets up grids based on bars and beats. Now that the software knows how fast we are playing, it will set up the grid lines accordingly. The grid lines help us shift parts exactly to the start of measures and so on, and make editing easy. Setting up the tempo like this will make life very easy for us down the road. Also, when recording future tracks, Cubase can play its click track at the right speed!

Adding in the loop is as simple as dragging it from the disk to the playing field and dropping it on bar one (the far left) (**FIGURE 16-3**).

FIGURE 16-3   ▼ **Drum loop added**   *Screenshot used by permission of Steinberg Media Technologies.*

Looking at **FIGURE 16-3**, we can see the loop is now on audio track one and is eight bars long. We know this because the ruler directly above the drum loop is set to show bars of 4/4 time. The ruler is numbered bar by bar, so the loop stretches to the eighth bar visually.

Eight bars are not long enough for this song, so we can easily duplicate the loop, looping it for as long as we need. This is done easily via the duplicate command. We will duplicate the file eight times for a total of sixty-four bars, which should be plenty long for us.

Now that the loop is long enough, let's start adding some other tracks.

Adding Bass

Using an audio interface and a microphone, we can add in our bass part. All we have to do is tell Cubase what input we are plugging the microphone into, set our record levels properly, and press record. **FIGURE 16-4** shows the input for the bass on audio track two. It's set for the "IN 1," which is where the microphone is plugged into on our particular audio interface.

FIGURE 16-4

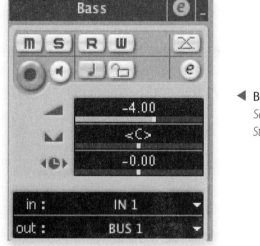

◀ Bass input

Screenshot used by permission of Steinberg Media Technologies.

Now that we've set up the channel and our level is full and not distorting, we simply record the part. This part will also be a looping section, because the recorded bass part is only four bars long. Once we record the part, we duplicate it the same way we duplicated the drum loop so that the bass plays until the end of the piece (the sixty-fourth bar). Let's take a look at what we've done so far (see **FIGURE 16-5**). We now have some nice backgrounds to work with. Let's move on to some MIDI parts.

FIGURE 16-5   ▼ **Drums and bass tracks**   *Screenshot used by permission of Steinberg Media Technologies.*

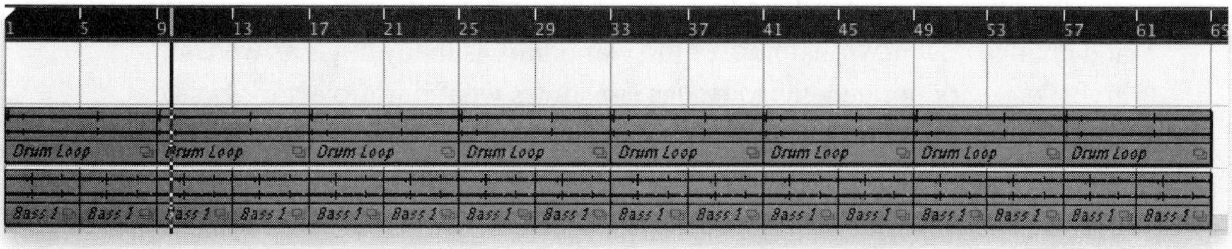

Adding MIDI

The next track to add is some piano. Since we don't have access to a real piano, MIDI will have to do for now. Using a MIDI keyboard and a virtual instrument sampler, we can add a realistic MIDI piano. Assuming your MIDI interface is hooked up and your keyboard is attached with MIDI cables, all we need to know is what channel your keyboard is talking on. In this case, it's channel one, which is usual for keyboards. We simply tell Cubase to listen for input on channel one. As for the output, we tell Cubase to route the MIDI output into MachFive, a sampler by MOTU equipped with a great piano sound. Look at **FIGURE 16-6** to see how we hooked the signal together inside the computer.

FIGURE 16-6

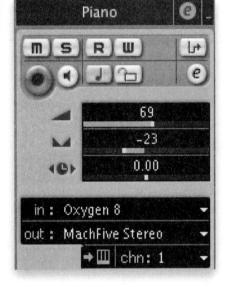

◀ MIDI piano setup
Screenshot used by permission of Steinberg Media Technologies.

A quick test of the keyboard and we have sound. We rewind the project and hit record to add some MIDI piano. After recording the piano chords, the MIDI tracks show up in our song. Instead of waves of sound, you see blocks of notes; this is how MIDI is represented in Cubase. We can go back later and change any individual note of the piano part as many times as we want. In this case, it's not necessary. **FIGURE 16-7** shows what our project looks like now.

FIGURE 16-7 ▼ Piano, bass, and drums *Screenshot used by permission of Steinberg Media Technologies.*

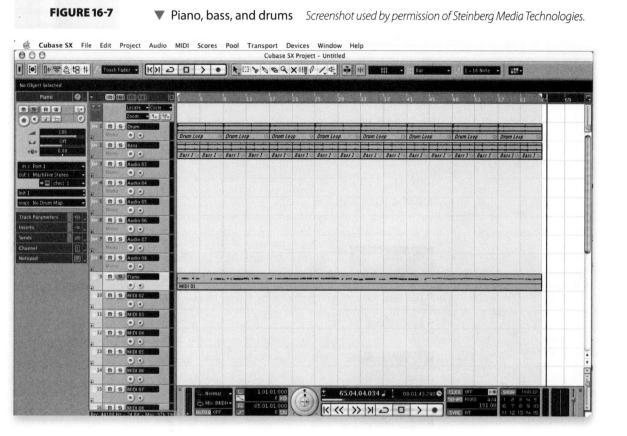

MIDI Quantize

For MIDI tracks, no command is more powerful than the quantize command. Our piano part was played well but suffers from occasional rhythmic inconsistencies. Some parts are slightly early, others slightly late. This is not serious enough to warrant rerecording the entire part. It's also a bit of overkill to go through each note, one by one, and change when they occur (although you could easily do that with Cubase's MIDI editing). The quantize command allows you to nudge the notes back and forth in time automatically so they play in tempo better. This is the most important reason to set up the tempo before starting. If Cubase doesn't know how fast our song is, it can't apply the quantize effect properly.

To apply the quantize effect, all we do is highlight the MIDI part, pull down the MIDI menu, and select quantize. Poof! Instant rhythmic perfection.

Building

Repeating the steps, we add two more audio tracks of guitar, one vocal track, and another MIDI track of a string part. To clean things up, we remove the empty tracks to improve the look and feel of the program. **FIGURE 16-8** shows what our song looks like now.

FIGURE 16-8   ▼ Full arrangement   *Screenshot used by permission of Steinberg Media Technologies.*

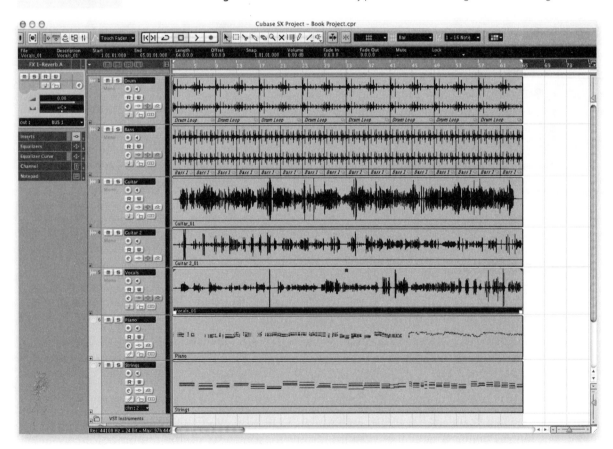

Mixing

Now that the tracks for our song are recorded, we need to mix the separate signals to get a good blend. We can do this inside Cubase, via its virtual mixer. A quick press of the F3 command launches this mixer (**FIGURE 16-9**).

FIGURE 16-9   ▼ Cubase mixer   *Screenshot used by permission of Steinberg Media Technologies.*

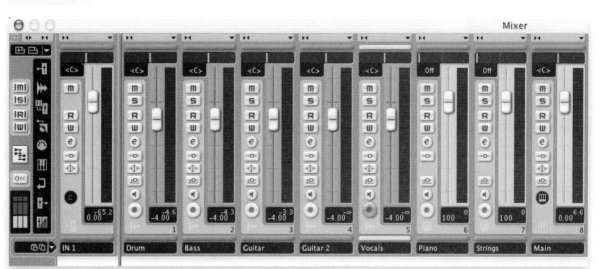

Each track is labeled by name, and the mixer gives you the immediate control of volume and pan for each track. Volume is handled by the vertical slider with the white button, while pan is at the top of each track, via a horizontal bar that you can drag from the left to the right. Setting the pan directly in the middle (called center pan) equally distributes the signal to both speakers. Dragging the bar to either side assigns the pan toward either the left or the right speaker.

Through some trial and error and patience, we mixed the basic volume of each track to our liking and set the pans. **FIGURE 16-10** shows what it looks like when we are finished.

FIGURE 16-10   ▼ Setup mixer   *Screenshot used by permission of Steinberg Media Technologies.*

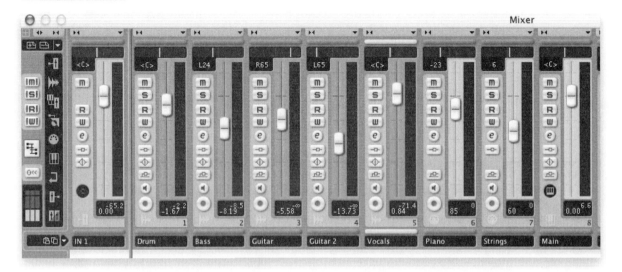

Adding Effects

To sweeten the mix, we need to add some effects to the individual tracks. Let's work with the vocal track, which could use some light compression and reverb. We launch the audio channel settings for the vocal track by clicking the E on the vocal track in the mixer. Hitting the E launches the window shown in **FIGURE 16-11**.

This channel-settings window includes slots for inset effects, a full graphic EQ section, the send controls for mixing in auxiliary effects, and another copy of the volume and pan adjustments we can adjust. This is where we add the compressor and the reverb to the track. We can also apply some EQ to soften up the sound. **FIGURE 16-12** shows the finished window after adding a compressor and reverb and EQ.

On the left side, we added the insert effect of a compressor (we set it for light compression to even out the track). On the right side, we routed the signal into the reverb effects channel to add some lush reverb to the vocals. Last, in the center we manipulated the EQ to take some of the high end away to soften the vocal track.

The result here is a great-sounding vocal track. Since every track will need varying amounts of EQ and effects, it's pointless to discuss exactly what we

FIGURE 16-11   ▼ Audio channel settings   *Screenshot used by permission of Steinberg Media Technologies.*

FIGURE 16-12   ▼ Final vocal channel settings   *Screenshot used by permission of Steinberg Media Technologies.*

used in this example. Your situation will be different. But we will say this: we used a soft compressor with a low ratio of 2.5:1 to even out the track. Any more would have been overkill here.

FIGURE 16-13   ▼ **Large mixer**   *Screenshot used by permission of Steinberg Media Technologies.*

Repeat

We went through each track and added some effects and EQ to the sounds that needed it. We did this in the exact same way we mixed the vocal track, using the audio channel editor that Cubase provides for every separate track. So now, after some tinkering, we have a nice-sounding mix. And we did it all inside the computer, using nothing more than a mouse and the included tools. Pretty cool!

FIGURE 16-13 shows an alternate view of the mixer after we have made our adjustments; this time we configured the mixer to show EQ above the volume faders. This is just one way to configure the mixer—you could easily show effects, inserts, and basically anything else you wanted. This is just another way that software is flexible for you.

MIDI Editing

The string part contained a small mistake. What we originally planned as a simple C triad (C-E-G) came out as C-F-G#—oops! Think the part is ruined? Think again. Simply double-click on the string part and the arrange window opens up a piano roll editor. The piano roll editor shows notes in the form of a graph. The up-and-down movement of the line signifies the levels of pitch being played. As the line moves left to right, it indicates the location of the levels within the bar and the duration of the notes. **FIGURE 16-14** shows the offending chord in the piano roll editor.

FIGURE 16-14

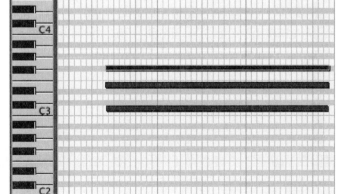

◀ MIDI piano roll editor (before)
Screenshot used by permission of Steinberg Media Technologies.

You see the piano keyboard vertically on the left side. The corresponding horizontal lines are the notes. The longer the horizontal line, the longer the note. For this example, we need to move the note F down to E and the note G# down to G. This is as simple as selecting the notes (the horizontal bars) and clicking down with the arrow key. Not only does the note move down, but you can also hear where it's going while you do it. Two simple key clicks and our notes are perfect again. **FIGURE 16-15** shows the completed edit.

FIGURE 16-15

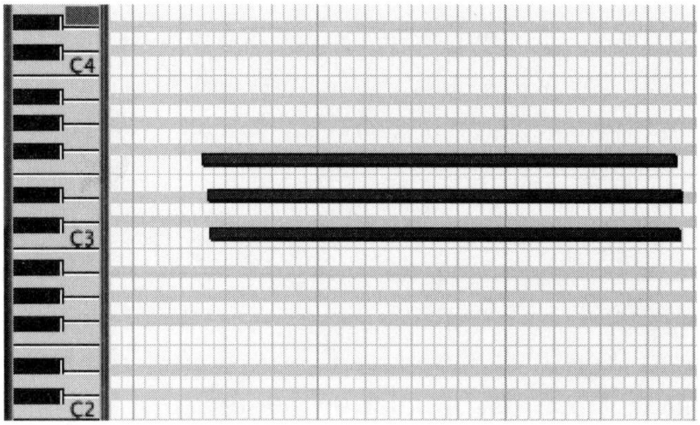

◀ MIDI piano roll
editor (after)
*Screenshot used
by permission of
Steinberg Media
Technologies.*

Bouncing

Now that we have something we like, the last step is to bounce this to a file we can burn or make into an MP3. We do it in a few simple steps. First, we have to tell Cubase where the song starts and ends. We do this using song position markers found in the ruler. We simply drag the right handle out to the last bar of the work. **FIGURE 16-16** shows the time markers set correctly for our masterpiece.

FIGURE 16-16   ▼ Audio markers   *Screenshot used by permission of Steinberg Media Technologies.*

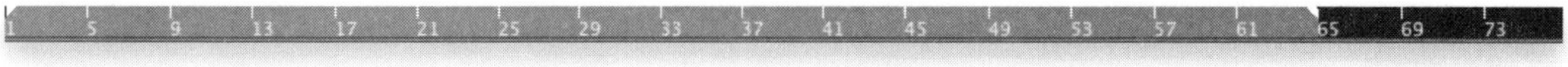

Now that the arrangement is defined, we can go ahead and bounce the work into a file suitable for burning or posting on the Internet. To do this, we pull down the file menu to the export command and select audio mixdown. We get the mix-down window shown in **FIGURE 16-17** where we see an impressive number of choices. Now take a look at our completed mix-down window (**FIGURE 16-18**) ready for mix down.

FIGURE 16-17

◀ Mix-down window (before)
Screenshot used by permission of Steinberg Media Technologies.

FIGURE 16-18

◀ Mix-down window (after)
Screenshot used by permission of Steinberg Media Technologies.

Let's hit upon the important things we changed: We made sure the channels were set to stereo, the resolution was set to 16 bit, and the sample rate was set to 44.100kHz. These are the CD standard settings for audio files. Failure to set the controls properly will yield some interesting results, not to mention it probably won't burn. If it does, the audio will be very slow, or very fast. Once you have those settings down, press the save button and you're good to go. Remember to name your file and look where the computer will save the file to so you can find it later. Now you can burn it in the software of your choice.

What's next? We've barely scratched the surface of what Cubase can do. Our simple demo song just got us started. You can add more instruments, layer track upon track, add more MIDI tracks, more loops—you name it! Some of the topics you might want to explore are more complicated audio editing, applying the other built-in effects available, bussing, punching in and out, automating volume and effect parameters, and adding more virtual instruments. The list goes on and on. As you can see, it wasn't that hard to get up and running thanks to the intuitive interface and powerful tools. With a little bit of imagination and know-how, you can create fantastic music.

Chapter 17

The Future of Home Recording

If you think what's available on the home recording market isn't already amazing, wait till you hear what's next. To say it's the future is a bit of a misnomer because it's already here. In a few years, all the technologies described in this chapter will be as common as the 4-track recorder. If you're looking to take your home studio to the next level, use what you read about here as your road map.

Many Multichannels

One of the limitations on home recording has always been not enough inputs and outputs. Whether it's a studio-in-a-box system or a full-blown computer interface, the downside has always been the limited number of inputs. It's not too hard to find an eight-channel computer-recording interface, but what if you need more than eight? One of the simplest ways to get more channels is to take advantage of digital inputs. Let's use the MOTU 828MKII as an example. The 828 boasts ten analog connections and ten additional *digital* channels. How do you take advantage of the extra inputs? On the 828 and many other recording interfaces, you will find an ADAT optical connection that sends its signal, not as analog audio, but as a digital signal, on a beam of light no less. Yes, fiber optics!

There is a new breed of microphone preamps that output to ADAT digital, which is also commonly referred to as "lightpipe." With a preamp like this, you can easily add eight additional microphone and line channels. Even better news is you can get a good unit for $200, and a *really* nice one for around $800. Voilà, more channels—better yet, more microphone channels, and you can never have too many of those.

FACT

ADAT lightpipe inputs are becoming common on recording interfaces. The addition of the lightpipe input might be a good reason to choose one interface over another—especially if you expect to grow with your studio.

The other alternative is to look for an interface that allows "daisy chaining." Daisy chaining is a way for interfaces to flow into one another, combining the inputs. Many of the MOTU interfaces allow you to chain as many together as you want. If one interface has sixteen inputs, then buying a second one and linking it together gives you thirty-two, and so on. If you were recording a jazz big band, for instance, thirty-two channels would come in handy.

Higher Quality

All commercially available CDs are recorded at 16bit/44.1kHz sample rate. Sample rate refers to how many times the analog wave is scanned per second when it's converted to digital information. At 44.1kHz, the audio wave is scanned 44,100 times every second. While that might seem like a lot, it's possible to have a higher sample rate. The highest sample rate available now is 192kHz, which means that the wave is scanned 192,000 times per second! The higher the sample rate, the better the computer can get a look at the incoming audio and the better it can make a representation of the true sound.

Sample rate is analogous to dots per inch on a printer. The more dots per inch, the sharper the printed image. The higher the sample rate, the better quality the audio sounds.

Sixteen-bit refers to the "bit depth," that is, the amplitude or loudness of a signal. At 16 bits, the incoming audio can have 65,536 levels of amplitude. If the incoming audio falls between one of the bits, the computer will round it up or down to whatever is closest. As a result, you're not getting a perfectly true picture of the sound. Higher bit depths are available to correct this. Twenty-four-bit allows 16,777,216 levels of amplitude, which as you can guess gives you a more accurate signal.

The new thing in home studio recording is improved audio quality via higher sample rates and bit depth. Sixteen-bit/44.1kHz is perfectly acceptable (and all CDs are encoded that way), but most audio programs accept higher sample rates and bit depth. You'll see 24bit/96kHz advertised, which means it's recording at a higher audio quality. That's the good news. The bad news is, not all software is capable of playing back high-rate audio, and neither can all interfaces. In effect, you can use a high-rate audio program and a 16-bit sound card and never reap the benefit. So be forewarned. Fortunately, most of the current audio cards allow 24bit/96kHz recording, and all the major software packages listed in Chapter 6 work at that rate as well. And yes, you can hear a difference.

Surround Sound

"Surround sound" is fast becoming a buzzword. If you have a home theatre system and a DVD player that supports 5.1-channel surround sound, then you've experienced just how cool surround sound is. The premise of surround sound is to expand upon the traditional stereo two-speaker setup. By placing five speakers and one subwoofer in a circle around the listener, audio can be panned in all directions around you.

Imagine being able to have music literally envelop you—as if you were sitting on stage with sound coming to you from all around. Surround-sound mixing is already present in Cubase, Sonar, Digital Performer, and Logic. When DVD audio becomes a standard reality, the home studio market will be able to mix sounds not only left and right, but back and forth and all around. When you hear it, it will blow you away.

Power

Computer software relies on the host computer processor for its power. The faster the machine, the more the software can do. This is almost always the limiting factor in the number of audio tracks and effects you can run at once. The next move is the introduction of computers with so much raw power that nothing you can throw at them makes 'em sweat. Imagine 100 audio tracks, each track with various effects. This has always been an issue for home studios: How many tracks and effects can you use before the computer starts to choke?

The day has finally come when you can create the music you want, regardless of the limitations imposed on you by your computer. Do you believe you'll never use fifty tracks in your music? You might be surprised that many of the "pro" albums that you know and love used 60 to 150 tracks of audio! Who knows what's around the corner, but one thing is for sure—as more power becomes available, developers will find creative ways to eat it up, which in turn will help you enrich the quality of your music.

If you have a recent computer that seems to get bogged down by your audio projects, consider using the new breed of PCI (peripheral component interconnect) and Firewire interfaces. They are a great way to give you a boost of power. These cards contain extra DSP (digital signal processing) chips to take the brunt of the audio effects off your CPU. If you have a slower computer that can't keep up, adding one of these power-saving cards and offloading the processing to the card can breathe new life into an old machine. Universal Audio manufactures the UAD-1 card while T.C. Electronic makes the Powercore card. Both cards do the audio processing for you without affecting the computer's CPU. They also come with special plug-ins specifically built for those cards. Since Powercore comes in a Firewire version, even laptop studios can have power to spare.

FACT

Universal Audio has a long reputation of making some of the finest hardware audio gear around. Its UAD-1 is Universal Audio's attempt at digitizing some of their most famous products for use in a computer system.

Advanced Computers

The power of the home computer is driving up the possibilities of what we can accomplish in a computer-recording studio. We now have the ability to work with multiple monitors, large data drives, control surfaces, and even our laptops.

Double Monitor Monsters

Let's face it, your computer recording software has a lot of windows to look at. All the multiple views of information are what make computer systems so powerful; no one is complaining. But wouldn't it be nice to have a few of your favorite screens open at all times? All at full size? Get another monitor, or even get three! By adding additional monitors, you'll be able to keep your most-used windows open at all times, greatly enhancing your workflow. You can keep the mixer window open on one screen and the main arrangement window open on another for easy access.

FACT

Instead of purchasing multiple monitors, you could upgrade to a single large monitor (20 inches or above), which will allow you to keep a few windows open at the same time.

You will need a special video card that allows multiple monitor support, but these video cards are available on the market at a relatively low cost. With the popularity of LCD monitors, you can pick up a second tube monitor for next to nothing, and if you're really feeling generous, grab a third LCD monitor to round out the studio. It's definitely a luxury, but if you spend a great deal of time in your studio working on music staring at a small screen, it's worthwhile.

Bigger Is Better

When it comes to hard drive space, bigger is definitely better. Audio data takes up a lot of disk space. If you opt to record at higher sample rates, such as 24bit/96kHz, your music will take up even more space. Once you fill a drive, you're stuck. Either you have to burn sessions to a CD or delete some of your work. A much better idea is to buy a very large drive. Get the biggest disk you can afford and never worry about running out of room again.

ALERT!

When recording audio, it's a good idea to have a second hard disk dedicated to audio files only. You'll get better performance out of your computer that way.

Laptop Studios

Desktop computers were always faster than laptops, usually significantly faster. Audio interfaces mainly used PCI interfaces, and laptops were too small to fit PCI cards. For the longest time, laptops were for convenience, not power, but now they're just as powerful as their counterpart desktop computers. With the audio interface industry adopting Firewire and USB over

PCI, laptops are finally able to serve as recording studios. One of the biggest drawbacks to a computer studio has been the lack of portability. Laptops are changing all that. Fully featured multitrack audio studios are showing up in laptops everywhere. If portability and space are a big concern for you, a laptop might be the way to go.

Control Surface

When you mix on a computer, your mouse is the interface to the onscreen virtual buttons and faders. While the mouse might be a blessing for high-resolution editing, many find the mouse cumbersome and limiting when it comes to mixing. Anyone who's used to a mixing board misses the freedom to drag multiple faders in opposing directions and hit multiple buttons at once, none of which a mouse will allow. For those who are looking for a better way to interact with their systems, enter the control surface.

FIGURE 17-1

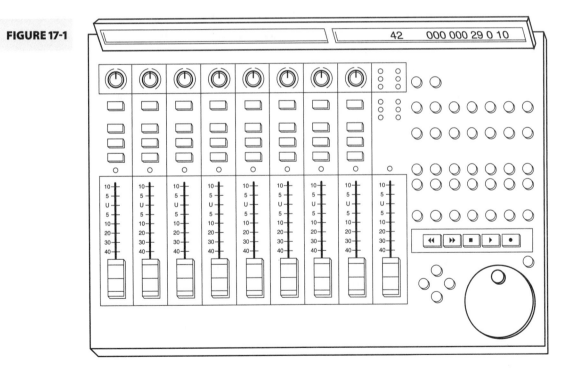

▲ Control surface

A control surface is a pseudo mixing board designed specifically for a computer recording setup. As you can see in **FIGURE 17-1**, the surface has faders for volume, mute and solo buttons, and a full transport section for controlling stop, start, and record functions. Moving a fader or pressing a button on the control surface initiates a matching command in the software. Those who are used to old-fashioned mixing will enjoy working this way. It's a much more efficient way to mix. Once you've tried it, you might never go back to the mouse again.

Totally Virtual

In the music industry, virtual is the name of the game and unlike virtual reality, virtual instruments are here, now. It all started with the synthesizer, which has been around for a long time, longer than most people know. The first synthesizers created sound, not with acoustic instruments, but purely by "synthesizing" or creating sound electronically. The early synths were analog in that they created sound with a series of tone generators and filters. They didn't have menus and editing screens. You turned the knobs until you got the sound you wanted. There were no presets either. Synths eventually became digital in that they created digital sound and manipulated the sound digitally through digital signal processing or DSP. It was only a matter of time before the digital processing came to the computer.

Virtual Instruments

A virtual instrument is a synthesizer or sound generator that is driven by software and lives within the computer. It's often played on a keyboard, and keyboard players love gadgets, especially extra sounds. Most of the modern keyboard workstations come equipped with literally thousands of sounds, and even *more* sound modules can easily be connected via MIDI. It's not uncommon for keyboard players to invest a great deal of money on additional sounds and filling floor-to-ceiling racks. This is fine for the studio, but taking it on the road is a big problem.

Just as plug-in effects have liberated the audio market, virtual instruments are revolutionizing music creation. A virtual instrument is simply a computer plug-in that responds to MIDI input, similar to how a sound module works in hardware; only these exist entirely in the computer. You don't have to be a keyboard player to use a virtual instrument, either. You can write MIDI parts with a mouse in your favorite software and the virtual instrument will do the rest. You can also compose with standard pitched notation if your software supports it.

FACT

Instead of touring with multiple keyboards and sound modules, jazz legend Herbie Hancock tours with a keyboard controller and an Apple laptop. All of the synth sounds are virtual instruments driven from software on his computer. Virtual instruments have replaced many of his vintage keyboards.

Virtual instruments usually come in a few flavors: true synthetic instruments, samplers (we talk about them in the next section), and re-creations of existing hardware. The virtual instrument market is growing by leaps and bounds. Every day something new is introduced to the market. Let's look at two different virtual instruments so you can get an idea of what we're talking about.

Absynth

Native Instruments makes some of the best virtual instruments around. Absynth is one of them, and it takes the idea of modular synthesis to the next level. You start with a basic waveform and you can apply filters and different oscillators to mutate the sound into whatever you please. You can do this to multiple sound sources and even sampling. You can do more than just basic shaping; you can edit every part of the sound in the most microscopic of ways. Absynth is a true synthesizer designed to create sounds and textures never heard before. While you could use it to re-create a flute sound, that's not what this kind of instrument is designed for (although it does have a nice flute preset). Take a look at Absynth's window (**FIGURE 17-2**) showing the waveform editing of a sound that we created. If you're into making sounds no one has ever heard before, Absynth is hard to beat. You'll find Absynth in use in music and film scores around the world.

FIGURE 17-2   ▼ Native Instruments' Absynth   *Screenshot used by permission of Native Instruments.*

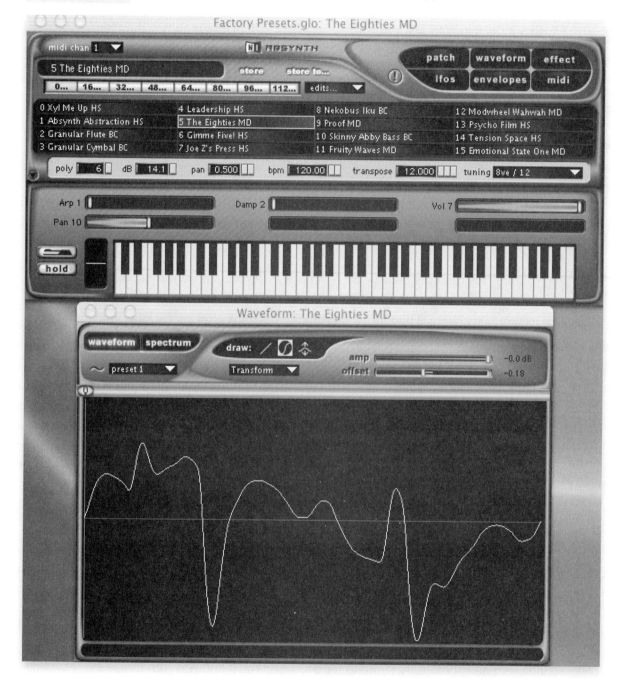

Trilogy

Trilogy, by Spectrasonics, is a total bass module. This plug-in does electric bass, upright bass, and combinations of those sounds. It captures the sound of a bass in such stunning realism that you'll have a hard time believing a real bass player isn't in your room. Like all virtual instruments, Trilogy responds to MIDI input, but it has particular sensitivity to volume and attack. Play a soft touch on your keyboard and Trilogy responds as if you had plucked the string lightly. Play harder on the upright bass sound, and you hear some of the characteristic "slap" of the string against the fingerboard that is so vital to the sound of a real bass. Take a look at Trilogy's main window (**FIGURE 17-3**). It gives you access to the different preset sounds built into Trilogy. In addition, you can adjust parameters of the sound, attack, and feel of how the instrument responds.

FIGURE 17-3 ▼ Spectrasonics' Trilogy *Screenshot used by permission of Spectrasonics.*

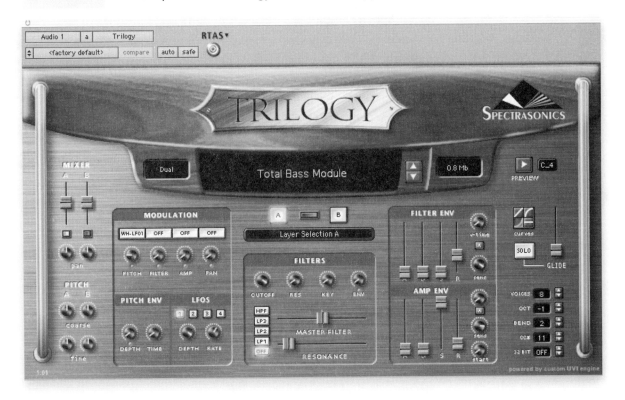

Virtual instruments are included in many recording packages such as Digital Performer, Logic, Cubase, and Sonar. Extra instruments can be purchased in a variety of plug-in formats to integrate with the recording software you own.

Samples and Samplers

The sample and sampler have been around for quite a while, but the concept is not new. First, what's a sample? Think of it this way: No matter how hard you try to re-create the sound of an acoustic instrument synthetically, it's never as good as the real thing. So, instead of trying to re-create the sound, you can record each note of a real acoustic instrument as an audio file and place each one in a machine capable of assigning MIDI notes to those sampled sounds. The sampled sounds, which are *real* recordings of *real* instruments, can be played back from a MIDI keyboard.

Now, what's a sampler? A sampler is a device that plays back prerecording audio via MIDI commands. For example, you play a note on your keyboard controller, and the corresponding audio sample is played back in whatever sound you choose. With this invention, musicians could play an authentic sounding trumpet from an ordinary keyboard. For a long time, rack-mounted samplers that read floppy disk and CD-ROM samples were important and frequently used studio tools. Samplers were also integrated into many of the high-end keyboard workstations available.

FACT

A great example of a sample that we all know and love is the slap bass heard throughout the *Seinfeld* TV show. That was a sampled bass, not a real one.

There are many virtual samplers made today. Some are standalone applications, such as the popular Gigasampler. Many are plug-ins that act as virtual instruments. EXS24 by Emagic, Sampletank by IK Multimedia, Kompakt and Kontakt by Native Instruments, and MachFive by MOTU are all examples of software-based samplers that exist as virtual instrument plug-ins.

As time marches on, the sound of the sampler is constantly improving. In the early days, the size of the audio files was prohibitive to "real" sampling. Usually, a few notes were sampled truly, and the rest were pitch-shifted to accommodate the full range. They sounded very good, but they were never replacements for the real thing. With the power of the personal computer and the affordability of larger hard disks, the quality of the sampler is hard to believe. **FIGURE 17-4** shows MachFive's main sampler window.

FIGURE 17-4   ▼ MOTU MachFive   *Screenshot used by permission of MOTU.*

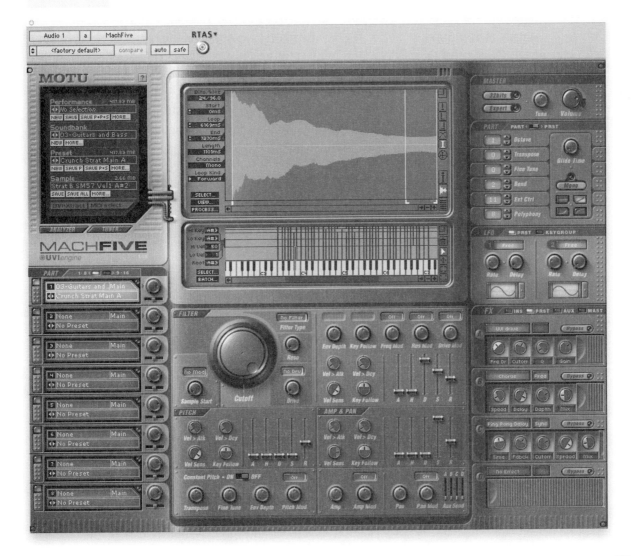

As you can see, you have a lot of control. Not only can you change aspects of the sample, you can sample instruments yourself! MachFive will also import and read most of the sampler formats available on the market, a great idea if you worked with a hardware sampler in the past and already have samples you like.

Cutting-Edge Software

There's some wild stuff available for making music these days. Let's take a look at two pieces of software that go beyond the normal realms of how music is traditionally made. Their names are Live and Reason.

Not everyone considers a home studio a place to record a band. Many musicians enjoy creating synthetic music with virtual instruments and synthesizers as much as traditional audio recordings of acoustic instruments. The home studio experience encompasses all forms and facets of music making.

Live

Live, written by Ableton, is considered a sequencing instrument. Live originally dealt only with audio—no MIDI, just audio. With the introduction of Live 4, Live is now a robust music-making application that includes MIDI and virtual instruments. Live uses a technique called "elastic audio" in which you tell Live the tempo of each section of audio and it takes care of lining them up together. If you don't know the tempo, Live figures it out for you. Live makes audio as malleable as MIDI does. For example, with MIDI, you can easily speed up and slow down parts; Live extends this power to audio. Live also lets you work with audio clips and samples and organize them into music on the fly—hence the name "Live." Many artists create music in real time on gigs with Live and a laptop computer.

Imagine being able to improvise clips at different tempos and to create arrangements on the fly. This is the power of Live. Live works as a standalone application, however its output can be routed into your favorite audio recording program through a process called "rewiring," allowing you to compose music in Live and pump the final sound into the software of your choice. **FIGURE 17-5** is a shot of Live in action.

FIGURE 17-5   ▼ Ableton Live   *Screenshot used by permission of Ableton.*

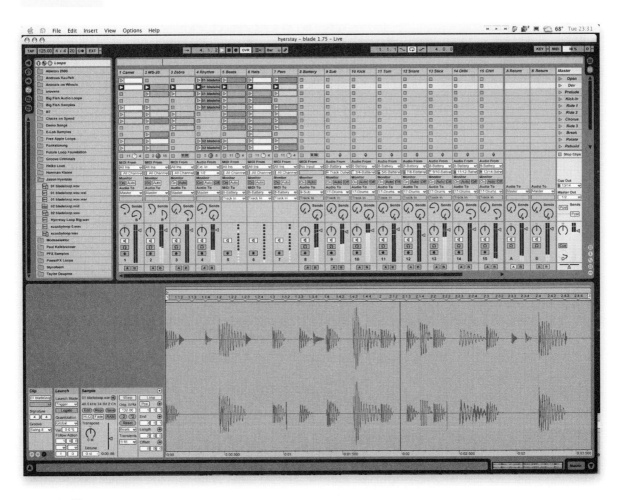

Reason

Reason software, written by Propellerhead Software, is a virtual synth studio. Reason has no audio capabilities; you can't record a note of audio into it. Reason is a self-contained set of every virtual instrument or synth you could ever think of needing. Synthesizers, drum machines, samplers, and effects processors are all bundled into Reason. Even the most computer-shy novice can start creating music in Reason. This fact has led to Reason's incredible popularity in the music world; everyone seems to have a copy these days.

Reason consists of a sequencer and virtual synth sections. You compose via MIDI and assign the tracks to the different virtual components that Reason comes with. Third-party companies have extended Reason's capabilities via "refills," which are extra sound sets and synthesizers for Reason. Just like Live, Reason's output can be piped into your favorite audio recording software via "rewire." **FIGURE 17-6** is a shot of Reason in action.

If you're looking for a new spin on your music, check out Live and Reason or any of the other cutting-edge products available on the market now.

FIGURE 17-6   ▼ Propellerhead's Reason   *Screenshot used by permission of Propellerhead Software.*

Chapter 18

Advanced Recording Tips and Techniques

At this point in your journey, you've gained a lot of general experience and have a good foundation to build upon. Your music sounds good. Your recordings are clean; there are no problems, yet they don't sound quite as amazing as you had hoped. What's missing? Every experienced engineer employs some tricks of the trade now and again to his or her own benefit. Here's a collection of those tips and tricks.

Multiple Microphones

Deciding on what microphone to use can be a difficult task. If your studio is large enough to hire a helper, it's not a problem to check microphone after microphone while you sit in the control room and listen back. But let's get back to reality! You don't have a big studio, and chances are it's just you working there. While you could set up a microphone, do a short recording, and repeat this several times with other microphones, that's inefficient and makes it hard for you to compare each microphone to each other. But how can you get around it?

If you have a few microphone inputs available, set up as many microphones as you feel like (or have available) in front of your sound source. Let's use a guitar amplifier as an example. Set up four microphones on the speaker, each at varying distances and locations around the speaker and record them all at the same time. When you play back the tracks, listen to them one at a time (use the solo feature) and see what you like best. If you find a microphone that stands out as the winner, delete the other tracks and you're done. Or you might also find that blending several microphones at once gives you the sound you're looking for. In any case, keep track of which microphone goes to which input!

Surgical EQ

The art of equalization is a subtle one. An experienced engineer can listen to a sound with such acute listening skills that he or she knows just what button to turn to get the sound perfect. Some of us aren't quite that skilled. Here's a great trick to find the frequencies that need help. (This trick works well on snare drums to find the frequency on the annoying ring that most snare drums exhibit.)

On a parametric or graphic (not a three-band) EQ, set the Q—the width of the EQ—as high as possible. This lets you pinpoint a very small range of frequencies, allowing a precise cut. Now set the level control all the way up as high as it will go, giving a maximum boost. Slowly change the frequency control, sweeping from the low to the high frequency. When you find the frequency that is responsible for the ring, or any other sound problem, you'll hear it easily. Because your EQ is boosted as high as it will go, that sound will jump right out of the speaker. Once you find the culprit, reduce the gain until the offending sound goes away. Repeat this as many times as you need to achieve the sound you desire. This is an absolutely essential trick as you learn about proper ways to EQ.

FACT

On a three-band EQ, you typically have no control over which frequency is boosted or attenuated. The three bands are set by the manufacturer and sweep through a general range of frequencies. This doesn't mean you can't use a three-band EQ effectively. If the frequency you need is built into one of the bands, than you're set.

Thicken Up Vocals

If your vocals seem to lack body and fullness, you might like this next tip. Duplicate the track so you have a second copy of the original, which is simple on a computer and a studio-in-a-box. On one of the copies, add a slight chorus and set the mix control to 100 percent full wet mix, so you hear only the effected sound. On that same copy, reduce the overall level of the track and pan it in the opposite direction of the original track. Placing the second copy in the other speaker will make the overall sound large and wide. You can substitute a short delay for a chorus effect for similar results. This technique of duplicating vocal tracks was used on many Beatles recordings.

Side Chains

Many effects utilize something called side chains, which are extra inputs. These are either physical inputs for hardware effects or virtual inputs for plug-ins. A side chain listens for the audio present in the side chain input and uses that information to trigger some part of the effect. The best example of this is used in radio broadcasting, called a "ducker": A compressor is applied after the CD player's output. The side chain inputs are from the announcer's vocals. When the announcer starts to talk, the compressor kicks in, lowering the signal of the music and letting the voice come through clearly. The music "ducks" under the voice.

Side chains like this are great for controlling guitar and vocal interaction. In most pop music, guitar is the focus when there isn't a vocal present. Using a compression side chain on a guitar track can help the guitar get out of the way when the vocal is present. Simply route the vocal track into the side chain inputs of the guitar's compressor. When the vocal is present, the guitar lowers slightly, and rises back up again after the vocal drops down. This can save you from having to manually "ride" the volume of the guitar track.

ALERT!

On a piece of hardware, side chaining involves patching with audio cables. With a plug-in, side chaining is done virtually by the computer. You set the inputs and outputs and the computer routes the audio for you.

Abusing Compression

Compression just might be the most important effect to master, but it's also easily abused. Compression is the reducing of the dynamic range of audio. Dynamic range is a natural part of sound. When you manipulate it, you are changing a big part of the sound. If done wrong, it can sound unnatural and harsh.

Breathing and Pumping

These are two common compression no-no's, and both are the results of overcompressing. Let's talk about how you can avoid them. Pumping typically occurs with bass drums, which have sudden loud bursts and can get too loud. Compression can help tame this, but if the kick drum is only one of many sounds fed into a compressor, you might have a problem. The bass drum will kick the volume up, and the compressor will respond by turning everything down with it, suddenly. Since the kick drum is rhythmic, the effect is noticeable because the volume pumps up and down in rhythm. The easiest way around this is to put the kick drum on its own separate compressor, or lengthen the attack time on the compressor.

Breathing is a side effect that occurs when vocals are overcompressed. If the compression is set too high, there is little difference in volume between the loudest sung phrases, and the quietest inhale. Breathing occurs when inhales and exhales of air are as loud as the rest of the sung parts. Not the nicest sound. You can get around this by setting the ratio lower and adjusting the threshold so that the quiet parts stay quieter, and the loud parts get squeezed slightly.

Overcompression: Hitting the Wall

Compression is the limiting of dynamic range. Dynamics are also referred to as nuance. If you overcompress a final mix, making the whole song very loud and not allowing it to dip down and have nuance, your music will suffer. NO ONE LIKES TO READ SENTENCES WITH ALL CAPS. IT'S ACTUALLY ANNOYING. That is a dramatic example, but that's what a listener hears when your mix is overcompressed with no dynamic range. This is also very prevalent in dance/house/club–style music. Don't let your music fall victim to the current trend of louder is better—it's not always so. Subtle compression will help a mix sit and sound nice, but overcompression will kill it.

On any compressor, the ratio control sets the amount of compression. The ratio of 10:1 is considered a very high ratio. It might work for some instruments, but anything over 6:1 is considered heavy compression. If you are experiencing breathing or pumping, look at your ratio first.

Reverb Mixes

Reverb is a great effect, and so necessary for just about all situations. It's also easy to get carried away with because reverb sounds so nice. Here is a very easy way to set reverb well. Start by applying a reverb on an auxiliary channel. Start mixing in the reverb until you can hear it's there. Then, back down slightly. It seems easy, but if you can hear an effect, you are probably overusing it. Dial it in, and then back off a touch.

Chapter 19

Recording a Demo and Putting It to Work

Recording a demo is one of the major reasons to invest in a home studio. Many of you probably got into home recording specifically for this reason. So, as a culmination of all of your experience and knowledge, it's time to get to work on a demo recording. After your demo is complete, we'll put it to work for you—gigs and recording contracts await!

So You Want to Make a Record?

It's time to stop fooling around—you need a demo tape or CD in order to further your career, get gigs, and spread your music to a wider audience. At some point it became clear to you that you can do this yourself, at home. For the cost of one professional-level demo, you can start building the studio of your dreams and record as many demos as you want.

Now that you've learned some techniques for recording and honed your skills, it's time to work on the details of the process.

Getting Organized

Organization skills are critical to getting anything done—not just on a demo. Too many groups just decide to start a demo without thinking it out, and eight months later they still aren't done. You should have a clear plan of what you're going to do and how you're going to do it. Here is a list of things to consider:

- What material will you include?
- What purpose will the demo fulfill?
- Are you equipped well enough to handle the project?
- Are you well enough rehearsed?

These are just some of the things to think over. Consider this: If you were going into a pro studio at an hourly rate, you'd go in prepared and well rehearsed so you wouldn't waste your money. Don't treat yourself and your home studio any differently; strive to be as productive as you can be.

Being serious and productive in your home studio does not mean that you shouldn't have any fun. The pressure-free environment of recording at home is a major plus.

Setting Up a Session

Plan your dates and session times in advance. Treat this just like anything else—have a goal for when you'd like to finish. A lot of this process depends on the size of your group, the complexity of your music, and what kind of studio you own. Bottom line: Approach the sessions just as you would in a pay studio. Too many times bands encounter the "we can do it next week" thinking and things keep getting put off later and later. Just because you aren't paying someone else per hour, don't let that change the efficiency and drive behind your own work.

Picking a Convenient Spot

Where will you record the demo? You might own the equipment, but your living conditions might not be suitable for live drums or loud guitars. Maybe you know someone who has a large basement you could use. Finding an adequate spot is key; the fewer compromises the better. If you're really in a pinch for space, you might be able to rent rehearsal spaces that are well soundproofed and that fit the bill. Sometimes opting to pay for a larger room in order to record the full band all at once vs. playing one track at a time in your basement might be worth the extra expense. Your demo will be your personal calling card and your key to many doors. Opt for the scenario that makes you sound better, even at a cost.

Rehearse, Rehearse, Rehearse!

Entering the studio, no matter what level studio, should be the last step in a long process of preparation and rehearsals. There is *nothing* worse than wasting time because you have a "loose" idea of what you're doing. The studio is usually a place to capture a final product. The only exception to this is when you go into your home studio specifically to write music together and record it. Many bands work this way—most with the luxury of booking eight months in a studio and having the recording company foot the bill. The other exception is to try recording your rehearsals. You might get a good enough take during a rehearsal to make into a demo.

The Fine Details

Depending on your situation, the details of recording a demo will be different from case to case. Variables like the number of players, space limitations, and equipment all play a role in deciding how you'll go about accomplishing your task. Here are some common scenarios and ways to work with them to the fullest.

Solo Demos

If you're working by yourself, it's a little easier to get organized because there are no other schedules to organize and generally you work when you want to. As most recording sessions do, you probably should start with a drum track. If your music doesn't need drums, find the track that has the strongest rhythmic elements, such as a strummed acoustic guitar, for example. Recording those elements first will give you a strong rhythmic base to lay the other tracks on top of. Once your rhythm tracks are down, there's no set order to how you should record the remaining tracks, although many people like to record the vocals last. If you use the studio as a compositional tool and write as you go, you might go about this process differently, and that's just fine—whatever works for you.

FACT

There's no set way to make a record. Everyone works differently—find out what works best for you and work it to your advantage.

Group Demos

Group demos require a little more planning because you'll likely have more technical hurdles and variables to deal with. One such hurdle is equipment. Suppose you picked up a home studio for yourself and joined a band later. Your current gear might not be the best for group work. The big question is, will you want the band to play together, or do you want to do it track by track, one player at a time? A few things influence this decision. Do you have the space for a live demo? For many, fitting four or more players plus live drums into a recording space isn't a possibility. Even if you do have the space, can you capture instruments well enough to minimize bleed-through from microphone to microphone? And does your recording machine allow enough simultaneous inputs? As we talked about earlier in the book, once you start recording a full band, you enter the "big leagues" of home recording, and the equipment that accommodates this can be expensive. You might be forced to record one or a few instruments at a time, building up a multitrack arrangement if your gear is limited in this fashion. Not to worry, many big-name albums are produced this way by choice, not by limitation.

Multitracking One Player at a Time

If you need to record one player at a time, there are a few things you can do to help you start smoothly. First, record the drums. You will thank yourself for having a rhythm track down first. Bass guitar usually comes next. Then, you can record the guitar and keys, finally adding in the vocals. The hardest thing about recording a group track by track is a certain feeling of disconnection when you split a group up like this. After all, you don't rehearse this way! Some of the magic and interaction between players will be lost. It's hard to get a vibe and a feel going when you break up the instruments, but the feeling of disconnection can be overcome.

When recording any direct instrument—one that doesn't use a microphone—nothing says you can't have the rest of the band playing along. Since there are no microphones involved, you won't hear the other instruments on the recording. Having the other players there might help the feel tremendously.

On the bright side, one player performing at a time means the overdubs and mistakes can be fixed with much less hassle.

Using a Click Track

Do yourself a favor and record with a click track, especially if you multitrack your demos. Live bands can usually regulate a good tempo together, especially one that breathes naturally, but when you take that element away, you might need a click to hold things together. Not to mention that it will help lock everything together anyway!

If you have drums, use the click track when recording the drums. Once that track is done, the click is no longer necessary because the drums act as a click from that point on.

Being Productive

Making the most of what you have, especially your time, is crucial. Here are a few pointers to get you going along the way. First of all, have a plan. Know exactly what you're going to record, when you're going to do it, and in what order things need to be accomplished. The more complex your band and recording situation, the more work you need to put into these preliminary steps. It might seem like overkill, but constructing a solid battle plan will make things go much smoother for all involved. You'll most likely have more fun and have a better final product, too!

Keeping to a Schedule

Try your best to set a realistic schedule for how long you think it will take to record each element of your demo. Just because multitrack audio allows you to go back as many times as you want doesn't necessarily mean that you should. Set time limits for various parts of your sessions to help the recording move forward. For example, set aside an afternoon for the guitar solos and promise yourself that whatever sounds the best that day is what gets to be on the demo. Period. Remember, it's a *demo*.

Budgeting

What will this cost you? Maybe nothing. If you have a home studio capable of doing what you want and you have the space and time, this might be a no-brainer. But for many people, extra costs might be involved—additional microphones, renting out ample space for recording, and so on. Budgeting is something that everyone in the project should share equally. You might have invested in the recording machine alone, but everyone involved can help split some of the additional costs such as studio time, additional microphones, and any other accessories that are needed.

Studio Log Sheets

Keep a log of each session you do and document details such as what microphones were used on what instruments, approximate microphone placements, microphone preamp settings, effect settings, and any other pertinent information you might need later. If you need to go back to fix or change something, you can get the sound the same by using the exact same variables for each instrument listed on your log. Log sheets like this are standard in pro studios. It's just a good idea to keep track of what's going on from song to song.

Keep Everything

There's no such thing as a bad take. A mistake here and there is no reason to delete anything. If you track to a computer system, most systems save all of your old work anyway. Even so, it's a good idea to save everything that you record because you never know what you're going to need later on. Especially in the case of multitrack recording, you might be able to assemble a "super" take, combining performances into one perfect take from past performances. And, if you ever make it big, you'll have your first "bonus" materials on your first greatest hits album: alternate takes!

Finalizing

At this point, you've completed the tracking stages and are ready to finalize the project through mixing and mastering at home or in a professional mastering house. It's time to turn your demo into a reality.

If you're working alone, mix to your heart's content. However, it's always a good idea to get some fresh ears on your work every once in a while.

Opinions do count here. If you're working in a band situation, don't mix alone, even if you have the equipment. It's a band project and everyone's opinions count. Listening to others ensures that you get everyone's input and help. The more ears on a mix, the better it will most likely come out. Once you've decided on a final mix candidate, burn out a few copies and listen to the work on as many different systems as possible. Try cars, home stereos, computers, Discmans, iPods—you name it. If it sounds good and clean, you might be able to put it out for the world to hear. If it doesn't, go back to the studio to even out the rough parts.

When mixing for extended periods of time, it's important to pace yourself and take breaks. Ear fatigue can make mixing very difficult. For every hour you mix, take a fifteen-minute break.

In recording, mastering is the last stage in album production. For a demo, however, you might not need it. Mastering generally deals with loudness, balance, and song sequence. If you've done a good job with levels during your recording, you might not need any loudness maximizing to make your signal strong enough. Depending on the purpose of the demo, mastering might not be worth the extra money. If the demo is not destined for release and is solely for solicitation of work and record contracts, mastering might not be necessary. If you feel that you need it, try mastering at home. There are many mastering plug-ins available for the computer-recording world.

Optimal Sound Quality

How does it sound? Good? Really good? What, exactly, sounds good? Is it just that you captured a good performance, or did you create a sound that stands on its own? If this demo is going to serve you well, it should sound as great as it can. You saved a ton of money recording it at home, and you also kept all of your creative control. Now's the time to ask the hard questions. Did you mix it as well as someone else could? Does it need to be mastered? Did this demo turn out to be album quality after all? If you answer yes to any of these questions, you want to investigate some options.

If you intend to distribute your recordings, gain radio airplay, or establish music industry ties, your recording needs to sound professional. Professional sound doesn't mean just mixing, noise, balance and effects, but more important, loudness. Loudness is one of the most critical parts of the mastering process, and it's one that should not be overlooked. We've all been frustrated by commercials that play too loudly on TV, forcing you to lower the volume—and then, *bam!* The next show comes on normal volume and you can't hear a thing. It's happened to all of us. That rogue commercial wasn't properly mastered.

Imagine that your demo gets into the hands of a club owner who takes the time to listen to your work. If he or she can't hear it properly due to mixing and mastering issues, then you might have closed that door. It might make sense to take your work to a professional mixing and mastering studio to help you put the finishing touches on your work. If something doesn't sound right, and you're not quite sure how to fix it, investigate some professional options. You'll also learn a lot.

Don't let obvious mistakes remain part of your recording. No matter what your original intent was, this demo has taken on a deeper purpose if you've decided to go public. Even if you started out wanting to simply utilize your new gear and you got a better result than you expected, polish it as best you can. Fix the mistakes your public will hear. If you intend to sell this recording, your demo is your personal calling card. Obvious mistakes in your recording ring out as clearly as spelling errors in a typed resume. You never know who's going to listen to it.

E ALERT!

Like diamonds, recordings are forever! That quick two-song demo you gave out at your first gig might appear on the Internet after you have hit it big. It would be a shame to have a poor performance follow you around.

Define Your Purpose

When you're working on recorded music, it's really important to understand that there are different standards for different purposes. What's this for? Is it just to get a few local gigs? To get a record deal? If you're looking for a local gig, you might be able to get away with slight imperfections and rough edges in the recordings. But if you're looking for a record deal, then the stakes change and the rules are different.

Getting Gigs

If you're looking to score some gigs, you need to do your homework. When looking into clubs, find out what kinds of crowds and what sorts of people frequent the establishments. You need to ask yourself, "What will this owner or booking agent want to hear?" While it would be wonderful to believe the owner is into music for music's sake, you're not being realistic. Live music plays one role—to make the establishment money. In turn, you are compensated. Owners want to hear music that fits into their normal mold, and they rarely take chances on new formats and risky groups. With that in mind, you need to customize your demo to address the most important need for each potential client. What that means is putting what you consider your best and strongest material for a particular client on the first track of the tape or CD. Chances are, the client won't even listen to the second track. It's very common to put together multiple demos with different track orders and song content for different purposes. One size might not fit all in this case.

ALERT!

When handing out a CD for a prospective gig, make sure to place your contact info and phone number as many places as possible. Make sure to mark the CD itself, because CD cases often get separated from their contents.

On the first track, which might get only one minute of play time, the club owner simply wants to hear that you aren't a joke. He or she most likely won't listen very carefully to your content. The club owner wants to make sure you fit into the mold so that regular customers are happy with your sound and new customers are drawn in by it.

Getting a Record Deal

If you're an ambitious sort, you might want to seek a record deal. After all, your home studio has allowed you to produce quality content just for this purpose. While sound quality is always important, it really comes into play if you want to get a record deal. The person hearing your music most often is involved with music production. Chances are, he or she listens to music all day long from bands that either go the home studio route or pay top dollar for professional demos. If your recording sticks out with poor quality, soft levels, and other anomalies, you might be dismissed right off the bat. That's why getting it right the first time is ideal—you might never get a second chance.

Making Good Connections

Making blind submissions (sending your recording to a club or company you've never contacted or made a connection with) is as good as sending your demo to the trash. Most record companies on principal throw out all blind submissions due to the large number of submissions they receive—you aren't the only one who wants a record deal, you know! Club owners often act the same way. They open the envelope, see a CD they don't recognize, and toss it into the trash. You have to make some connections beforehand.

Talk to the Right People

In the case of clubs and other performance venues, a simple phone call beforehand will usually suffice: let them know who you are and ask if they are accepting submissions. They might respond with questions about your music and the audience you typically play for—this is usual. After you make the initial contact, send your demo by mail or preferably show up in person and place your demo right in their hands. After that, back off. Give them time to listen, and then follow up a few days later.

Making recording industry contacts is much more difficult. The sheer number of people who try to initiate contact with artist relations persons makes this part difficult. You might be very hard-pressed to get anyone on the phone at all. Most companies will tell you they don't accept unsolicited submissions, which is a nice way of saying: Don't call us; we'll call you. How can you overcome these difficulties? Many times agents and managers can make these connections for you. If you're serious about getting into the industry, seeking the help of an agent or manager could open doors. Really good agents and managers already have connections and close ties with record companies, and their submissions often have a chance of getting noticed.

QUESTION?

How can I get in touch with agents and managers in my area?
Ask other artists and bands with whom they've worked and had good experiences. For every good agent, there are ten agents who won't help your chances of success. Start with someone you know has had successes.

Fill a Need

The other aspect of doing your homework is understanding what "they" want. "They" can be record companies, clubs, venues, and concert promoters . . . you name it. The name of the game is filling a need. If you don't read this part correctly, then your demo will end up in the trash again. When looking into prospective avenues to distribute your demo, make sure you're close to what "they" are looking for. For example, you're wasting your time and money sending a heavy metal demo to a jazz club. The same goes for record companies—make sure you've got a similar style to other artists on the company's current roster. Otherwise, the company won't take your demo seriously and you'll end up simply wasting your money.

Be Professional

The presentation of your demo makes a big difference, so package your demo in a great-looking press kit. A basic, professional-looking press kit includes these elements:

- **A CD:** These days, the format of choice is the compact disc. Tapes are falling out of favor.
- **A label with contact information and track listing:** Make sure you place a nice label on your CD—don't just write on it with a black marker! Label it using a very inexpensive adhesive CD labeling kit on your home computer. Include contact names, phone numbers, and track listings.
- **A group biography:** Include a well-written, typed bio of your group. The more information you provide the better. You might want to list places you've played, awards, and so on.
- **A picture (optional):** A picture helps to humanize the group and elevate a boring press kit into a human reality.

All of these elements should be packed either in a nice folder or bound together in some way that they don't get lost or separated.

No matter how good this demo of yours is, if you don't *act* professionally, you're not going to get very far. Anyone who's been involved in the recording industry will tell you that product is not always as important as personality. Getting your demo to work for you will invariably involve phone calls and mailings, and it's to your advantage to speak and write as professionally as you can. That means being polite and considerate during your phone calls, especially when talking to the people who book gigs for concerts and clubs. These people are often overwhelmed with other responsibilities, so they might appear to "blow you off." It's so important to keep a perspective here and exercise an extra bit of patience. Remember, you need them; they don't need you. There are other demos they can likely choose from, so make a good impression by being nice. When writing and e-mailing, make sure to use good English and practice your written communication skills. Just because it's becoming normal and ordinary not to capitalize words in e-mail and instant messenger doesn't mean it's correct. So don't take that habit with you when you correspond (e-mail or snail mail) with prospective clients. And please, please spell-check your work!

The more pleasant, easy, and fair you are to work with, the better. This will pay off in your search for contacts, in winning repeat gigs, and in finding success in putting your demo to work.

Copyright Protection

One thing we've left out of this discussion until now is how legally to protect your songs. If you're writing your own material, you own the intellectual property fully. But if you don't take the proper steps to protect yourself, you might get yourself in trouble, or worse yet, give away your music. Here's the lowdown on the copyright process and how to secure your rights.

From the minute you record music, you own it. The law protects you as the creator of your work. However, if things get messy later on, you might regret not having filed a formal copyright application with the U.S. government. It's easy and inexpensive. All you have to do is fill out a few forms and send a check for each work, which typically totals $30 for a full album. If someone tries to use your music in the future without your permission, you will have a much better footing in court if you choose to litigate for copyright infringement of intellectual property. It's so simple to do; you really should take a few minutes and do it.

FACT

For more information on how to obtain copyrights, go to ✎*www.loc. gov/copyright*, the Web site of the U.S. Copyright Office. Web sites and resources for other topics of interest to the home studio owner are listed in Appendix B in the back of this book.

The American Society of Composers, Authors, and Publishers (ASCAP) and Broadcast Music, Inc. (BMI) are two organizations that serve to protect copyrighted music and its writers and publishers. Typically, you will need to join both ASCAP and BMI after you sign a record contract; most likely someone at the recording company will take care of setting all this up for you. If your song gets played on the radio, the organization—ASCAP or BMI—would make sure you get paid the proper royalty. It's very rare to have music on the radio without a label behind you—not impossible, but not common.

Selling It

If you plan on selling your music yourself, you will enjoy a few benefits and face a few hurdles, too! The first benefit is, you keep all the money after you subtract the cost of making the product. Nowadays, you'll make more money per CD if you sell the music yourself than if a record company sells it for you and gives you a cut. The only difference is the volume of sales you can achieve with a company backing versus selling it yourself. In either case, before you sell your music, you have to get it ready for duplication.

Duplication

The simplest way to duplicate your music is to burn it at home on a CD burner. The CD burner can be either a standalone system or part of your computer. Blank CDs continue to plummet in price, so the cost of making CDs at home is an attractive option. You can even produce professional-looking, computer-printed labels for both the front of the CD, and insert materials for the jewel case. Considering the price of good-quality printers these days, you can yield some impressive results at home with a relatively small investment.

For those who want to leave the duplication to the professionals, you'll find many options. Usually, these duplication companies deal in large runs, say of 500 or more CDs at a time. While you can order fewer, the difference in price between 100 and 500 is small enough that most people opt for more. You submit either a burned CD or mastered DAT and the company takes care of the rest. You can either supply the art or pay to have it designed for you. Professionally made CDs look better than homemade ones because the duplication company uses better printers and it inks the CD labels onto the case rather than applying inexpensive adhesive labels. It also shrink-wraps the CDs in plastic. There are literally hundreds of places that duplicate and package CDs. You can find them by searching the Internet and local music magazines and papers.

Selling at Gigs

Now that you have a CD ready to sell, either from home or a duplication house, the most logical place to start selling it is at your gigs. You can generate a lot of sales and buzz at live shows. Having CDs for sale, especially if you tour around, is essential for promoting yourself, not to mention for making extra money. Most bands that sell CDs at gigs don't charge record-store prices, and this can be an attractive reason for an audience member to buy the CD.

Selling Online

This is a relatively new concept for musicians. You can start off by selling your CDs on your Web site (if you have one), but keep in mind that many people are leery of sending cash or checks to a stranger (you). Lately Web sites are popping up to handle CD sales for people just like you. Sites like *www.cdbaby.com* not only take orders for you, they set up a Web page with sound samples and handle secure credit card ordering. Web sites such as these charge you a few dollars of each sale for the cost of hosting and shipping your CDs. You get a weekly check for the money you make. It's a great way to be seen and heard, because these sites are popular on the Web.

The Web site *www.cdbaby.com* is hooked up with the Apple iTunes Music Store, so any music that you sell through CDBaby will also be available for sale through Apple's legal download service. This is a great way to spread your music.

Barcodes

If you plan to sell your CD online or through stores you'll need a barcode. CD manufacturing houses typically provide barcodes, but you can also get them yourself by registering for one (through a CD duplication facility). Whenever your CD is purchased, the barcode is read by "soundscan," which is a service that tracks record sales. If you're interested in having real sales data to show a record company, get all of your CDs equipped with barcodes.

Promotion

Promotion and the art of self-promotion are key. Here are some tips on how to get your music heard outside of your immediate area and circle of people:

- Give away your CD to clubs and restaurants—or any other place that plays music. It's a great way to get heard. With luck, it just might get played, and someone is bound to ask, "What was that?"
- Get your local record stores to stock your CD. It's great exposure.
- Start with your own Web site. What, you don't have one? Get one! It's now expected that you will have one. Get your CD and sound samples on there too.
- Post a page with a site that hosts songs for free. Such sites (*www.mp3.com* and *www.iuma.com* are two examples) are haunted regularly by many adventurous Internet music seekers looking for up-and-coming music. You get a simple page with your music, links to your site, and places to buy your CD. Many bands have gotten their start this way and you can too. Harness the power of the Internet.

Promotion is all about creating excitement and buzz about a product. The more work you do to create excitement for your project the better you will do. Use the examples above or try your own. In the end, success is the end result of a great deal of legwork and hard work. Promotion takes patience and diligence. You can do it!

Chapter 20

Making Money from Your Investment

What started out as a simple idea to record your own music has quickly turned into a full-blown home studio. You've invested in equipment, cables, microphones, and all of the other trimmings that go into a studio. More important, you've honed your chops as a studio engineer and are making some consistently high-quality recordings. It might be time to recoup your investment and open your home studio for business.

Use It or Lose It

We've all suffered from G.A.S.—gear acquisition syndrome—at one point in our lives. This ailment can strike suddenly, forcing you to buy gear. It's quite hard to stop, too! By this point you've probably built up a nice studio, and now all the pieces are in place for you to make great music. But it begs the question, how often are you going to record yourself? Will you feel guilty when thousands of dollars lay dormant in the corner of your room or basement collecting dust between creative lulls? You need to do something with it. If you've been working hard in your studio and sharing your work with your friends and colleagues, word of mouth will spread about what you're up to and before you know it, people will ask you to record their music. Since you enjoy the process of recording, you might take on a few small projects, usually for friends or friends of friends. Then it happens: You realize you're ready to open a studio—a project studio.

Buy as You Need

If you feel a little unprepared equipmentwise, especially in the microphone department, don't worry, most of us face this at one time or another. One of the best ways to build up your studio is to invest back into your studio the money you make from your recording sessions. For example, if you're working with a good singer and lack a good vocal microphone, use the money you earn on the sessions with that singer to buy a nice vocal microphone. (Actually, figure what you'll earn and invest in the microphone ahead of time, so the singer can use it!) Microphones are just one of the elements you might need to invest in for your home studio.

Microphones are one of the only pieces of equipment that haven't progressed dramatically in the past fifty years. Investing in a good set could last you fifty more years. Say that about other equipment!

Headphones and monitors usually need to be upgraded, especially if you work with large groups and you're used to going solo. You should invest in a few pairs of decent headphones and a headphone distributor, which is a device that takes one main output and splits the signal to multiple headphones. After you do a few diverse sessions and fill in the missing parts of your gear, you should be well set up to handle just about anything that comes your way.

Back It Up

"Be prepared" is the Scout motto, and it should be your recording motto as well. Be prepared for every foul-up possible. If you record to tape, have extras. Expect cables to break when you need them most. Batteries always run out at the most inopportune times! You should have a stock of extras of everything possible. While microphones are usually pretty durable, it's always a good idea to have some other choices in case something goes wrong.

Part of running a professional studio is being able to handle anything that comes your way. Murphy's Law will get you every time if you're not prepared for it: Anything that can go wrong will go wrong! Expect the most ridiculous errors! This is very true of the computer-recording world. Back up the sessions to CD or DVD often. Don't take chances losing valuable data.

ALERT!

If you run a computer studio, make sure you have the number of a computer tech on hand—you never know when you might need to call. And most likely, the computer will find a reason to crash on you when you really need it; (Don't sweat it; it happens to the best of us.)

Creating a Business Plan

If you're ready to step up to the plate and start offering professional services, you need some sort of plan. All businesses operate with a business plan that lays out the goals and objectives and, most important, how they intend to make a profit. For you, the studio owner, the goal is pretty simple: Earn money by recording music for clients. If you're ready to take the jump to the next level and open up a pro or semipro recording facility, there are a few things to keep in mind.

Opening Up Your Studio

It's time to open your doors to the public. But don't do it until you have a good understanding of what kinds of projects you can and can't handle. Space usually dictates this. Do you really have enough space for a full band to record without getting squished together? How about isolation booths and separation between instruments? How's your monitoring setup? Can you provide headphone mixes to each player? These are just a few of the things you'll be expected to offer your clients and customers.

A truly professional studio won't have many limitations. If you have doubts as to your ability to take on certain projects, be up-front and honest with prospective clients. You might rework your pricing to accommodate the unusual session. Most artists and clients won't mind a financial break to work out some issues, especially if the environment they work in (your studio) is a low-pressure environment. More and more people are recording in project studios like these because of the flexibility that the setup affords them.

Working with Clients

When you start operating a studio, there's an immediate paradigm shift: Instead of working for yourself and making your own rules, you now work for someone else. The client is the boss. While you might have expertise in the area of the technology and the art of recording, when it comes to the sessions, the client is in charge—it's the client's money. This is not to say that you're a slave! There are plenty of times when you can contribute your point of view and advice. For example, it's not uncommon for clients to come in expecting to work a certain way, and your experience can teach them that it won't work the way they thought. Just remember your role—you're there to ensure that whatever music they play is re-created in a flattering way. It's your job to set up all the equipment for optimal sound quality. What a client wants most in a facility is a trouble-free, low-stress environment. Be cool and calm, even if they go back and retrack the bass solo fifty-six times. Working well with others will spread your reputation quickly around town. The combination of high-quality sound, a relaxed environment, and flexibility will entice clients to come your way.

Compatibility

What do you record on? What do the "real" studios record on? Compatibility is a major issue these days. If you want to be a full-range studio, compatibility should be a concern of yours, too. It's really common to have a client come into the studio to track parts but not expect you to mix it. Is your software compatible with the client's? Or the client might have worked on a session at another facility and wants to add or rework parts at yours. Can you do it? Being able to interact with other studios and their equipment is another component of the services you can offer.

More studios use Pro Tools than anything else, but that doesn't mean it's the only thing anymore. Logic, Cubase, Sonar, and Digital Performer are found in studios all over. If you want to ensure the most compatibility with other studios, call every studio in your area and find out what they record with. You'll find that many of the older studios still offer good old analog tape, which you'll most likely never afford on top of digital recording systems. Some argue that analog sounds better than digital. For many now, it's a matter of price. Analog is more expensive to run and maintain. Even with all that, many still use analog for its distinct sound.

If you feel confident in keeping things in-house and not interacting or collaborating with other studios, feel free to use whatever recording system you are most used to.

Setting Rates

After you generate some initial interest and decide to take on a few beginning projects, you'll be faced with the eternal question, "How much will you charge?"

Give It Away; Give It Away Now

At first, you should probably charge nothing at all. But why on earth would you want to do this? Because you aren't a professional recording engineer . . . yet! The most valuable thing you lack is real work experience, so just like any business, you should set your fees based on experience. If you're just starting out with no experience, taking no money will ensure that someone will take a chance recording in your studio. What will they have to lose? Instead, you'll gain valuable job experience, which gets you closer to earning real money, and you'll also get a chance to figure out challenges brought on by other people's music.

FACT

By taking no compensation, you might feel more freedom to experiment with sounds and recording techniques that otherwise could have wasted a client's time and money.

As you get more comfortable handling diverse recording situations, you'll be able to start charging fees comfortably. There's nothing worse than charging by the hour and looking like a fool when you have to read the manual to figure out how to do something. Before you start charging, you should be able to run a full session without help from manuals, product guides, or calls to other home recording buddies.

Charging Fees

If the idea of giving away your services makes you uneasy, charging a flat fee might be a great way to go. It's also the next step after you stop giving away your services. Flat fees for projects are great when you're starting to record larger projects. However, just starting out, you lack the experience to give a fair estimate, or so you fear. If you recorded 100 bands this year, you have a pretty good idea what to expect. If not, give it your best guess. This is better than charging by the hour and worrying about the ticking clock over your shoulder. This will also help save face when you encounter the occasional setback like messing up a take or being unfamiliar with a software feature and having to call tech support!

When you're ready to charge by the hour, what do you want to charge? This is a hard question to answer. If you start too low, you'll gain a reputation as a "budget" studio and you'll find it hard to raise your rates later, especially if you've built up a client base. If you start too high you run the risk of turning away prospective clients. Find a few studios in your area that you believe are comparable to yours. Remember to judge with your ears and not with your eyes. It's possible now to have everything self-contained within the computer, so how your studio looks is irrelevant compared to the quality you can produce. If you can match the quality of other studios near you, you can match the rates. You can always lower a touch to attract new business and to help get your career rolling. Many studios charge anywhere from $40 to $100 per hour to record. You can find studios that are both cheaper and more expensive, but most of the smaller-level studios charge the same. Since every area has its own pricing, do some research and check around to find out what's considered "fair" in your area.

QUESTION?

How can I figure out where to set my prices?
Call other professional studios in your area and see what they charge per hour. You should charge less than a professional studio until you are experienced and time-tested. You might want to start with charging half of what other studios charge at first, raising your rates as you become more confident in your work.

Don't expect to make a fortune from recording, especially if your equipment and facilities aren't amazing. But you can make some nice pocket change and upgrade your equipment as you go along.

Working the Control Room

Regardless of the size of your studio, you will spend most of your time in the control room. If you have a small studio, this won't be a separate room, but more of an area where the main gear is. From your command post in the control room, you have one or two jobs to handle. The first is engineering duties. The other might be producing.

Engineering

The role of the engineer is to set up all the equipment and generally run the sessions. The engineer (you) places all the microphones, sets the levels, runs the mixing board, operates the recording equipment, and performs the mixing, editing, and whatever other duties come up. The client expects to take advantage of your knowledge of how all that works. In a typical engineering role, you have limited interaction with the music. You might be asked for opinions, but in many cases, the client will have a clear idea of what he or she wants from the sessions. You're there to help capture and preserve the music the way the client wants.

Producing

The role of the producer is varied in the recording industry. Basically, a producer is involved with bringing an idea of music to life. This might involve booking studio time, hiring players, arranging music, and so on. In the home studio, your production duties might come as a result of the artist's lack of concrete ideas. Or in many cases, an artist will come in with a basic framework and look to you and your studio to fill in the gaps. This can mean finding players to perform the music, or programming the music into a sequencer to provide a backdrop.

Don't assume that only garage rock bands come into studios. Electronic music, rap, and hip-hop rely heavily on producers to guide the sessions. It's not uncommon in a rap session for an artist to have only the rhyme committed to paper. The backdrop and beats are usually the collaborative responsibilities of both the producer and the artist. It's up to you as a producer to be familiar with all parts of your studio to help facilitate this. If you're set up only for recording sound, make that clear to all prospective clients.

Getting Known

Spreading the word that you operate a studio can be a tall order, especially if you aren't connected with the professional music scene in your area. Here are a few tips to help get you started.

Advertising

Nothing beats a little bit of good advertising. A well-placed flier can do wonders, so be sure to place fliers where musicians go. Music stores often provide space for this via a bulletin board. Also visit the local clubs and talk to bands and artists who perform there to let them know about what you're doing. You might even try advertising in print ads. Many cities have music papers that highlight local music, and they, like most the papers, rely on advertising revenue to keep afloat. Your advertisement can go a long way if it's printed where musicians look for these kinds of services.

FACT

When advertising, a picture or description of your gear might help attract customers. Make it sound exciting—don't just list the equipment you own. Describe what you can *do* with it all. Think about how to make it sound appealing and inspiring—what would make *you* want to record in your studio? Use that to attract others.

Flexibility

Nothing spreads faster than good news. In the recording studio world, news about a flexible studio and studio owner will spread like wildfire. Having a reputation for a clean, well-equipped studio with competitive rates and an easy, flexible working environment will be your greatest advertising tool. Remember, business of any kind is about people and the interaction with those people. The easier to deal with and more flexible you are, the better your reputation will be.

Allies

Get to know the other studios in your area. Make sure they know who you are, and what you're doing. Little things like directing a group you can't accommodate to another studio will generate goodwill from that studio, and one day it might return the favor and send a few clients your way. The more often you do this, the better reputation you will get among studio owners.

You might also want to consider bringing some of your best work to other studio owners. Regardless of how well set up your studio is, if another studio is going to refer business to you, it needs to be certain you can deliver a high-quality product. Nothing looks worse than recommending a studio that turns out a bad product. In those cases, neither studio ever gets a call again from the disappointed artist, and the bad news will travel—fast.

Word of Mouth

Most studios rely solely on word of mouth for new business. You'll find that the more clients who have good experiences in your studio, the better you will do. If your work is good, they will tell their friends. This is one of the reasons to keep prices low—it helps draw a steady stream of clients. Another way to drum up business is to ask for studio credit, another kind of word of mouth. If you record a demo for a band, make it part of their contract that they must list the name and number of your studio somewhere in the CD packaging. That way, if the CD sells a lot of copies, your number will be distributed to a lot of people.

Chapter 21
Next Steps

This has been quite a journey. What started out as a simple endeavor to get involved with home studios has grown into more than you expected. If you've developed a taste for recording and are thinking about getting further into the field, you're not alone. Many of the professional engineers got their start the exact same way. If you're looking to step up to the big league, read on.

Continuing to Grow

When you think back to how it all started, which might not have been that long ago, it's really quite amazing what you know now. If you've read and studied carefully, nothing in the studio should be unfamiliar to you. However, knowing what a compressor does, for example, and being able to use one effectively are two very different things. One thing is for sure—knowledge of gear and how it functions is the first step. The second step is using it and getting good results.

What the Pro Studios Use

In your own studio, you're the boss. You make every decision. When a vocalist comes in to record, you know just what microphone and preamp to use. You know because you own the gear—which is usually limited by your budget. But what would you do if you walked into a fully decked-out pro studio that owned every piece of gear and offered a virtually unlimited array of choices? If price were no object, what microphone would you choose? It's not uncommon for pro studios to own a few Neumann microphones, which typically go for $2,500 and more. You may never have heard these microphones, so how will you know when to use them? What you need to do is get some valuable in-studio experience. There are a number of ways you can go about this. The first is to connect with other studios in your area.

FACT

Renting gear is a great way to get to know better equipment and add professional sounds to your recordings without making a considerable investment. You can rent microphones, preamps, and outboard gear on a per-day basis.

Networking and Apprenticeship

You might already have strong connections with other home studio owners, but to gain experience in the professional world, you should be aware of the professional facilities around you. No matter where you live, there's bound to be a major studio somewhere nearby. Get on the phone and call the studio owner and see if he or she will meet with you. Explain what you are doing and that you would like to expand your skill set. Be sure to talk to the professional engineers there, too, to find out how they got their start. You'll find that most have a similar story—another experienced engineer took them under his or her wing and taught them the ropes. Maybe one day they can take you under their wing! It might also help your cause to bring some of your recorded materials along with you. These days, when it's so easy to own a home studio, in order to be taken seriously, you need to have some product to show for yourself. Perhaps, if you make a good impression, you will have opened a door for future use. If you can refer some work to the studio, either before or after the meeting, all the better for you. But don't stop with just one studio; make the rounds to others in your area. Networking is how you will branch out and succeed. It's really about who you know, as much as what you know.

Internships

Many people, in many fields, started off their career as an intern. Internships are the ground-floor entrance into many careers. An internship is basically a throwback to the apprenticeships of old. You donate your services for no, or virtually no, pay. In the recording industry, you do the "gofer jobs" like running cables, setting microphones and microphone stands, and grabbing coffee—basically anything they need you for. In return, you get free job training in a specialized area. You gain the all-important "experience." Many students take part in internships during their college years to boost their resumes. As an intern, you donate your time and services, and in turn you get to be around and learn how to use equipment that you probably would never have access to otherwise. Most important, you will be able to watch a master engineer at work. Nothing is more valuable than that!

As time rolls on, you will be able to do more. Studio control rooms are generally not that large, so it will be easy for you to observe everything going on. In time, as you prove yourself and gain a good relationship with the staff, you may even be offered a job there. If nothing else, you can walk away from the experience knowing that you have gained experience on professional-level equipment. That is worth the price of admission—your free time.

Affiliations

Becoming a member of a professional group is another way to get yourself plugged into the professional world. One of the most prominent groups for audio recording is the Audio Engineers Society (AES). Membership benefits include journals, publications, and access to a wealth of information and technical documents on the art and science of recording. Every year AES holds an annual trade show, usually in New York City, and as a member you're entitled to attend. The AES show is one of the largest recording and sound technology trade shows in the world. Major manufacturers display new products and conduct hands-on product learning and information workshops. The floor is teaming with professionals in the audio world. From authors to engineers, producers to postproduction, you can meet a cross section of the recording world. Conventions like these are a great way to learn about new technology and equipment.

Take a Class or Clinic

Just hanging around a studio may not get you the education you desire. You may not get enough time behind the console, learning to mix and master. Gaining in-studio experience can be obtained in other ways besides internships and jobs at recording studios. Recording is a subject that in the past twenty-five years has become a legitimate course of study. Some universities even award degrees in recording technology and engineering.

Many manufacturers realize that home recording is a growing and important field. To tap into this market, more and more companies are hitting the road to provide clinics and training on their respective products. More often than not, these clinics and classes are part of music store promotions. The good news for you is that they are usually completely free of charge. You'll be able to get some hands-on experience with new and exciting technology and gear. Manufacturers usually send their most knowledgeable people to demonstrate, so you will gain great insight watching them move around software and hardware they know like the backs of their hands. Many of these clinics are hands-on, so you will be able to try out the products as well.

ESSENTIAL

You can find the dates and locations of clinics and in-store events by visiting your local music store or looking online at gear manufacturers' Web sites.

While this may not be the learning experience you're looking for, you will get a chance to use cutting-edge tools. Since many stores hold clinics like this often, continually showing your face will help you make connections as well. If you do decide to buy something you see demonstrated, you might also receive a discount.

Get Certified

If you plan to record with software such as Pro Tools or Logic Pro, you should know that Digidesign (Pro Tools) and Apple (Logic Pro) offer certifications on their software. Getting certified requires that you attend a Pro Tools or Logic training facility, which can be found in major cities across the country. You enroll in courses that train you in the specifics of the application and how to be a master user. After you complete the various levels of certification, you are awarded a certificate as an operator and expert user. Typically, earning certification is no small feat; the tests are usually quite comprehensive. If you're awarded a certification, rest assured that you know and can use the software in practically every way!

After you complete the certification process, which, by the way, costs a moderate sum of money, you may have a better shot at landing a job. Many studio owners who honed their skills in the analog world aren't necessarily adept at computers and computer software, even if it's designed for recording. They will sometimes hire people specifically to operate the software aspect of the recording industry. Bigger studios and postproduction houses might employ a computer engineer, who is responsible for running the computer and the software.

For a prospective employer, certification doesn't mean that you know anything about making music in a studio. The certifications cover only the software you're trained on, which we all know is only one part of the equation.

You'll need other experience in general recording studios to help boost your resume. However, certification and experience make for an attractive resume. Considering the pace at which computers are changing the recording industry, certification in software may be a prerequisite one day.

FACT

While certification in music software may be new, other industries that rely on computers have used certification as job prerequisites for years.

Do Research

As you sift through the various written materials on recording and the science of sound, you're bound to get confused somewhere along the way. At first, the abundance of technical jargon may seem overwhelming, but it's an element that you need to understand. The science of sound is something worth taking the time to research. This could be as simple as taking a few books out of your local library on the study of acoustics and reading up. If you have little background in physics, you might want to start with some simpler texts explaining some of the simpler concepts that you will build on in the more advanced books.

The other area to study in great detail is electricity. Sound recording is the conversion of sound into electricity. Understanding electricity is vital to truly knowing your way around a studio and all its components. Many of the topics we brushed over, such as decibels, voltage, impedance, and polarity, are part of an engineer's breadth of knowledge. Getting deeper into these topics will help you out greatly. You won't pick up all of it in a studio; at some point everyone has to hit the books.

Research can also mean being informed of late-breaking technology and advancements in the field. Reading the journals and published magazines on recording and engineering will help you keep current and up-to-date with everything that's going on. The role of the engineer keeps changing, and the lines that used to separate artist from producer and engineer are continually blurring. Be as informed as possible.

FACT

Engineer Dave Moulton has produced a set of CDs called "Golden Ears." The CDs help you to learn EQ bands, compression, reverb settings, and so on by ear—the equivalent to ear training for musicians. It's a great set of tools to train your ears.

Go Back to School

As we had mentioned earlier, universities and specialized recording schools are popping up all over the place. Going back to school can be a wonderful way for you to get your chops together. All of the programs are based on the division between classroom learning, concept learning, and actual hands-on experience. There are two courses of study: a specialized recording certification program and recording course work within a music degree program.

Specialized recording certification programs are typically six to twelve months in length and focus completely on the recording process. You spend a great deal of time in a studio where your learning is almost all hands-on. You will walk out with enough experience to confidently start working in virtually any situation. You receive certification at the conclusion of the program.

QUESTION?

I have a day job—how can I find time to train for recording?
Many of the programs offer night and weekend courses for people in your situation. Look for programs that can fit into your schedule. Check with local schools for more information.

In programs that integrated recording into a music degree, you spend a large percentage of your time learning recording engineering. You also take the other music-related courses like theory and history that all music majors take. If you're a younger reader thinking about getting into music and recording, an integrated program like this might fit the bill. Best of all, you'll walk out with a bachelor's degree—and that means something in every field. If you're serious about getting into the professional recording industry, pursuing formal education might be just what you need.

What You Should Walk Away With

No tears, but we are at the end of our journey. Here's a checklist of topics and concepts that you should carry away from this book. If you have a good grasp on these concepts, you have built a solid foundation.

- **Basics of Sound:** You should know what makes up a sound. You should understand terms like amplitude, frequency, and pitch. The basics of how music is represented as waves of sound is very important.
- **Microphones:** What is the difference between a dynamic and a condenser microphone? When and where should you use each? How should you set them up for different instruments? What are the usual techniques involved?
- **Gain:** What is gain? What sorts of instruments output low gain? Why can't you plug a microphone directly into a recorder? Why won't a guitar or bass plug in directly without a direct box? How do decibels affect sound loudness? And why is 0dB considered the loudest possible sound in some cases, and the softest possible sound in other cases?
- **Mixers:** What is the basic idea of a mixer? When would you need one, and when might you be able to live without one? What's an insert? How do aux effects work? What is busing and what do master sections do?
- **Effects:** What's the difference between an insert effect and an auxiliary effect? Why would you never use reverb as an insert effect? Why would you only use compression as an insert effect? Why would you want to use compression? How are a compressor and a limiter related?
- **Software:** What is MIDI? What is a sequencer? What does DAW mean? What are the advantages of computers? Conversely, what are their disadvantages? What is the function of a plug-in? What is a virtual instrument?
- **Recorders:** What is a studio-in-a-box? How is digital sound different from analog? Why is tape dying out? Why is the computer so popular?

That's a lot of questions to answer. No one said this was going to be easy, but you persevered. Congratulations! Now what? Get into your new studio and make music. Mix and master it into a final product and get it out for the world to hear. Post it online. This whole journey started out with a desire to capture and create music. Don't lose sight of that goal. Enjoy your new studio.

Appendices

Appendix A

Recording Equipment Manufacturers and Suppliers

Appendix B

Additional Resources

Appendix A

Recording Equipment Manufacturers and Suppliers

Recording Equipment and Software

Alesis
✐*www.alesis.com*
Keyboards, effects, and recording devices

Antares Audio Technologies
✐*www.antarestech.com*
Auto-Tune and other plug-ins and hardware

Behringer
✐*www.behringer.com*
Full range of entry-level recording equipment

Brian Moore Guitars
✐*www.brianmooreguitars.com*
Guitars with built-in MIDI interfaces

Cakewalk
✐*www.cakewalk.com*
Sonar program

Digidesign
✐*www.digidesign.com*
Pro Tools software and hardware

Emagic
✐*www.emagic.de*
Logic program

Fostex
✐*www.fostex.com*
Recording equipment

IK Multimedia
✐*www.ikmultimedia.com*
Computer plug-ins (AmpliTube, SampleTank, T-Racks)

Lexicon
✐*www.lexicon.com*
Really nice hardware and software effects

Mackie
✐*www.mackie.com*
Mixers and control surfaces

Mark of the Unicorn
✐*www.motu.com*
Digital Performer, MIDI, and audio interfaces

M-Audio
✐*www.m-audio.com*
Manufactures and distributes everything from audio interfaces to software to MIDI keyboard controllers

Music Notation

Native Instruments
✐*www.nativeinstruments.com*
Virtual instruments and samplers

Presonus
✐*www.presonus.com*
Audio equipment for studio recording

Roland
✐*www.rolandus.com*

Makers of just about everything from keyboards to recording equipment!

Sibelius
✐*www.sibelius.com*
Tools for writing music symbols

Spectrasonics
✐*www.spectrasonics.net*
Best in virtual instruments

Steinberg
✐*www.steinberg.net*
Cubase, Nuendo, Wavelab, and other software

Tascam
✐*www.tascam.com*
Recording equipment

Universal Audio
✐*www.uaudio.com*
High-end recording equipment and the UAD-1 Card

Wavearts
✐*www.wavearts.com*
Computer-recording plug-ins

Waves
✐*www.waves.com*
Really nice computer-recording plug-ins

Yamaha
✐*www.yamaha.com*
Recording equipment

Microphones

AKG Microphones
✐*www.akg.com*

Electro-Voice Microphones
✐*www.electrovoice.com*

Neumann
✐*www.neumannusa.com*

Rode Microphones
✐*www.rode.com.au*

Sennheiser Microphones
✐*www.sennheiser.com*

Shure Microphones
✐*www.shure.com*

Studio Projects Microphones
✐*www.studioprojects.com*

Internet Sites to Shop

✐*www.audiomidi.com*
✐*www.musiciansfriend.com*

✐*www.samash.com*
✐*www.sweetwater.com*
✐*www.zzsounds.com*

CD Duplication Companies

✐*www.diskmakers.com*
✐*www.oasiscd.com*

Appendix B

Additional Resources

Magazines

Computer Music
✐*www.computermusic.co.uk*

Electronic Musician
✐*www.emusician.com*

EQ
✐*www.eqmag.com*

Mix
✐*www.mixonline.com*

Recording
✐*www.recordingmag.com*

Sound on Sound
✐*www.soundonsound.com*

Tape Op
✐*www.tapeop.com*

Web Sites

✐*www.ascap.com*
Music licensing
✐*www.bmi.com*
Music licensing
✐*www.cdbaby.com*
Sell your CD online.
✐*www.groups.yahoo.com*
Discussion groups for everything you can imagine, audio too!
✐*www.harmonycentral.com*
Music industry news
✐*www.homerecording.com*
An informative site
✐*www.loc.gov/copyright*
U.S. Copyright Office
✐*www.marcschonbrun.com*
Author's site
✐*www.osxaudio.com*
Everything Mac audio
✐*www.prorec.com*
Great reviews and articles

Books

Anderton, Craig. *Multieffects for Musicians.* New York: Music Sales Corp., 1995.
—. *Craig Anderton's Home Recording for Musicians.* New York: Music Sales Corp., 1996.
—. *MIDI for Musicians.* New York: Music Sales Corp., 1984.
—. *Audio Mastering (Quick Start).* Bremen, Germany: Wizoo., 2002.
Franz, David. *Producing in the Home Studio with Pro Tools.* Boston: Berklee Press, 2003.
Moulton, Dave. *Golden Ears Audio Eartraining Program,* 4 vol. (8 CDs and manual). Sherman Oaks, Calif.: KIQ Productions, 1994.
Woran, John M., and Alan P. Kefauver. *The New Recording Studio Handbook.* Plainview, N.Y.: ELAR Publishing, 1989.

Index

THE EVERYTHING SERIES!

BUSINESS

Everything® Business Planning Book
Everything® Coaching and Mentoring Book
Everything® Fundraising Book
Everything® Home-Based Business Book
Everything® Landlording Book
Everything® Leadership Book
Everything® Managing People Book
Everything® Negotiating Book
Everything® Online Business Book
Everything® Project Management Book
Everything® Robert's Rules Book, $7.95
Everything® Selling Book
Everything® Start Your Own Business Book
Everything® Time Management Book

COMPUTERS

Everything® Computer Book

COOKBOOKS

Everything® Barbecue Cookbook
Everything® Bartender's Book, $9.95
Everything® Chinese Cookbook
Everything® Chocolate Cookbook
Everything® Cookbook
Everything® Dessert Cookbook
Everything® Diabetes Cookbook
Everything® Fondue Cookbook
Everything® Grilling Cookbook
Everything® Holiday Cookbook
Everything® Indian Cookbook
Everything® Low-Carb Cookbook
Everything® Low-Fat High-Flavor Cookbook
Everything® Low-Salt Cookbook
Everything® Mediterranean Cookbook
Everything® Mexican Cookbook
Everything® One-Pot Cookbook
Everything® Pasta Cookbook
Everything® Quick Meals Cookbook
Everything® Slow Cooker Cookbook
Everything® Soup Cookbook

Everything® Thai Cookbook
Everything® Vegetarian Cookbook
Everything® Wine Book

HEALTH

Everything® Alzheimer's Book
Everything® Anti-Aging Book
Everything® Diabetes Book
Everything® Dieting Book
Everything® Hypnosis Book
Everything® Low Cholesterol Book
Everything® Massage Book
Everything® Menopause Book
Everything® Nutrition Book
Everything® Reflexology Book
Everything® Reiki Book
Everything® Stress Management Book
Everything® Vitamins, Minerals, and
            Nutritional Supplements Book

HISTORY

Everything® American Government Book
Everything® American History Book
Everything® Civil War Book
Everything® Irish History & Heritage Book
Everything® Mafia Book
Everything® Middle East Book

HOBBIES & GAMES

Everything® Bridge Book
Everything® Candlemaking Book
Everything® Card Games Book
Everything® Cartooning Book
Everything® Casino Gambling Book, 2nd Ed.
Everything® Chess Basics Book
Everything® Crossword and Puzzle Book
Everything® Crossword Challenge Book
Everything® Drawing Book
Everything® Digital Photography Book
Everything® Easy Crosswords Book
Everything® Family Tree Book

Everything® Games Book
Everything® Knitting Book
Everything® Magic Book
Everything® Motorcycle Book
Everything® Online Genealogy Book
Everything® Photography Book
Everything® Poker Strategy Book
Everything® Pool & Billiards Book
Everything® Quilting Book
Everything® Scrapbooking Book
Everything® Sewing Book
Everything® Soapmaking Book

HOME IMPROVEMENT

Everything® Feng Shui Book
Everything® Feng Shui Decluttering Book, $9.95
Everything® Fix-It Book
Everything® Homebuilding Book
Everything® Home Decorating Book
Everything® Landscaping Book
Everything® Lawn Care Book
Everything® Organize Your Home Book

EVERYTHING® KIDS' BOOKS

All titles are $6.95

Everything® Kids' Baseball Book, 3rd Ed.
Everything® Kids' Bible Trivia Book
Everything® Kids' Bugs Book
Everything® Kids' Christmas Puzzle
            & Activity Book
Everything® Kids' Cookbook
Everything® Kids' Halloween Puzzle
            & Activity Book
Everything® Kids' Hidden Pictures Book
            Everything® Kids' Joke Book
Everything® Kids' Knock Knock Book
Everything® Kids' Math Puzzles Book
Everything® Kids' Mazes Book
Everything® Kids' Money Book

All Everything® books are priced at $12.95 or $14.95, unless otherwise stated. Prices subject to change without notice.

Everything® Kids' Monsters Book
Everything® Kids' Nature Book
Everything® Kids' Puzzle Book
Everything® Kids' Riddles & Brain Teasers Book
Everything® Kids' Science Experiments Book
Everything® Kids' Soccer Book
Everything® Kids' Travel Activity Book

KIDS' STORY BOOKS

Everything® Bedtime Story Book
Everything® Bible Stories Book
Everything® Fairy Tales Book

LANGUAGE

Everything® Conversational Japanese Book
    (with CD), $19.95
Everything® Inglés Book
Everything® French Phrase Book, $9.95
Everything® Learning French Book
Everything® Learning German Book
Everything® Learning Italian Book
Everything® Learning Latin Book
Everything® Learning Spanish Book
Everything® Sign Language Book
Everything® Spanish Phrase Book, $9.95
Everything® Spanish Verb Book, $9.95

MUSIC

Everything® Drums Book (with CD), $19.95
Everything® Guitar Book
Everything® Home Recording Book
Everything® Playing Piano and Keyboards Book
Everything® Rock & Blues Guitar Book
    (with CD), $19.95
Everything® Songwriting Book

NEW AGE

Everything® Astrology Book
Everything® Dreams Book
Everything® Ghost Book
Everything® Love Signs Book, $9.95
Everything® Meditation Book
Everything® Numerology Book
Everything® Paganism Book
Everything® Palmistry Book
Everything® Psychic Book
Everything® Spells & Charms Book
Everything® Tarot Book
Everything® Wicca and Witchcraft Book

PARENTING

Everything® Baby Names Book
Everything® Baby Shower Book
Everything® Baby's First Food Book
Everything® Baby's First Year Book
Everything® Birthing Book
Everything® Breastfeeding Book
Everything® Father-to-Be Book
Everything® Get Ready for Baby Book
Everything® Getting Pregnant Book
Everything® Homeschooling Book
Everything® Parent's Guide to Children
    with Asperger's Syndrome
Everything® Parent's Guide to Children
    with Autism
Everything® Parent's Guide to Children
    with Dyslexia
Everything® Parent's Guide to Positive Discipline
Everything® Parent's Guide to Raising a
    Successful Child
Everything® Parenting a Teenager Book
Everything® Potty Training Book, $9.95
Everything® Pregnancy Book, 2nd Ed.
Everything® Pregnancy Fitness Book
Everything® Pregnancy Nutrition Book
Everything® Pregnancy Organizer, $15.00
Everything® Toddler Book
Everything® Tween Book

PERSONAL FINANCE

Everything® Budgeting Book
Everything® Get Out of Debt Book
Everything® Homebuying Book, 2nd Ed.
Everything® Homeselling Book
Everything® Investing Book
Everything® Online Business Book
Everything® Personal Finance Book
Everything® Personal Finance in Your
    20s & 30s Book
Everything® Real Estate Investing Book
Everything® Wills & Estate Planning Book

PETS

Everything® Cat Book
Everything® Dog Book
Everything® Dog Training and Tricks Book
Everything® Golden Retriever Book
Everything® Horse Book
Everything® Labrador Retriever Book
Everything® Poodle Book

Everything® Puppy Book
Everything® Rottweiler Book
Everything® Tropical Fish Book

REFERENCE

Everything® Car Care Book
Everything® Classical Mythology Book
Everything® Einstein Book
Everything® Etiquette Book
Everything® Great Thinkers Book
Everything® Philosophy Book
Everything® Psychology Book
Everything® Shakespeare Book
Everything® Toasts Book

RELIGION

Everything® Angels Book
Everything® Bible Book
Everything® Buddhism Book
Everything® Catholicism Book
Everything® Christianity Book
Everything® Jewish History & Heritage Book
Everything® Judaism Book
Everything® Koran Book
Everything® Prayer Book
Everything® Saints Book
Everything® Understanding Islam Book
Everything® World's Religions Book
Everything® Zen Book

SCHOOL & CAREERS

Everything® After College Book
Everything® Alternative Careers Book
Everything® College Survival Book
Everything® Cover Letter Book
Everything® Get-a-Job Book
Everything® Job Interview Book
Everything® New Teacher Book
Everything® Online Job Search Book
Everything® Personal Finance Book
Everything® Practice Interview Book
Everything® Resume Book, 2nd Ed.
Everything® Study Book

SELF-HELP/
RELATIONSHIPS

Everything® Dating Book
Everything® Divorce Book
Everything® Great Sex Book

All Everything® books are priced at $12.95 or $14.95, unless otherwise stated. Prices subject to change without notice.

Everything® Kama Sutra Book
Everything® Self-Esteem Book

SPORTS & FITNESS

Everything® Body Shaping Book
Everything® Fishing Book
Everything® Fly-Fishing Book
Everything® Golf Book
Everything® Golf Instruction Book
Everything® Knots Book
Everything® Pilates Book
Everything® Running Book
Everything® T'ai Chi and QiGong Book
Everything® Total Fitness Book
Everything® Weight Training Book
Everything® Yoga Book

TRAVEL

Everything® Family Guide to Hawaii
Everything® Family Guide to New York City,
    2nd Ed.

Everything® Family Guide to Washington D.C.,
    2nd Ed.
Everything® Family Guide to the Walt Disney
    World Resort®, Universal Studios®,
    and Greater Orlando, 4th Ed.
Everything® Guide to Las Vegas
Everything® Guide to New England
Everything® Travel Guide to the Disneyland
    Resort®, California Adventure®,
    Universal Studios®, and the
    Anaheim Area

WEDDINGS

Everything® Bachelorette Party Book, $9.95
Everything® Bridesmaid Book, $9.95
Everything® Creative Wedding Ideas Book
Everything® Elopement Book, $9.95
Everything® Father of the Bride Book, $9.95
Everything® Groom Book, $9.95
Everything® Jewish Wedding Book
Everything® Mother of the Bride Book, $9.95
Everything® Wedding Book, 3rd Ed.

Everything® Wedding Checklist, $7.95
Everything® Wedding Etiquette Book, $7.95
Everything® Wedding Organizer, $15.00
Everything® Wedding Shower Book, $7.95
Everything® Wedding Vows Book, $7.95
Everything® Weddings on a Budget Book, $9.95

WRITING

Everything® Creative Writing Book
Everything® Get Published Book
Everything® Grammar and Style Book
Everything® Grant Writing Book
Everything® Guide to Writing a Novel
Everything® Guide to Writing Children's Books
Everything® Screenwriting Book
Everything® Writing Well Book

Introducing an exceptional new line of beginner craft books from the Everything® series!

EVERYTHING
C·R·A·F·T·S

All titles are $14.95.

Everything® Crafts—Create Your Own Greeting Cards
1-59337-226-4
Everything® Crafts—Polymer Clay for Beginners
1-59337-230-2

Everything® Crafts—Rubberstamping Made Easy
1-59337-229-9
Everything® Crafts—Wedding Decorations
and Keepsakes
1-59337-227-2

Available wherever books are sold!
To order, call 800-872-5627, or visit us at *www.everything.com*
Everything® and everything.com® are registered trademarks of F+W Publications, Inc.